S0-AIT-352

Gender and the Language of Religion

Also by Allyson Jule

GENDER, PARTICIPATION AND SILENCE IN THE LANGUAGE CLASSROOM: Sh-Shushing the Girls

Gender and the Language of Religion

Edited by

Allyson Jule
University of Glamorgan, UK

#57564881

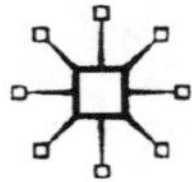

Editorial matter, selection and introduction
© Allyson Jule 2005
Preface and individual chapters © the authors 2005

All rights reserved. No reproduction, copy or transmission of this publication may be made without written permission.

No paragraph of this publication may be reproduced, copied or transmitted save with written permission or in accordance with the provisions of the Copyright, Designs and Patents Act 1988, or under the terms of any licence permitting limited copying issued by the Copyright Licensing Agency, 90 Tottenham Court Road, London W1T 4LP.

Any person who does any unauthorized act in relation to this publication may be liable to criminal prosecution and civil claims for damages.

The authors have asserted their rights to be identified as the authors of this work in accordance with the Copyright, Designs and Patents Act 1988.

First published in 2005 by
PALGRAVE MACMILLAN
Houndmills, Basingstoke, Hampshire RG21 6XS and
175 Fifth Avenue, New York, N.Y. 10010
Companies and representatives throughout the world.

PALGRAVE MACMILLAN is the global academic imprint of the Palgrave Macmillan division of St. Martin's Press, LLC and of Palgrave Macmillan Ltd. Macmillan® is a registered trademark in the United States, United Kingdom and other countries. Palgrave is a registered trademark in the European Union and other countries.

ISBN-13: 978–1–4039–4862–5 hardback
ISBN-10: 1–4039–4862–3 hardback

This book is printed on paper suitable for recycling and made from fully managed and sustained forest sources.

A catalogue record for this book is available from the British Library.

Library of Congress Cataloging-in-Publication Data

Gender and the language of religion / Allyson Jule, [editor].
p. cm.
Includes bibliographical references and index.
ISBN 1–4039–4862–3
1. Sex – Religious aspects. 2. Sex role – Religious aspects. I. Jule, Allyson, 1965–

BL65.S4G47 2005
200'.82—dc22 2005041532

10 9 8 7 6 5 4 3 2 1
14 13 12 11 10 09 08 07 06 05

Printed and bound in Great Britain by
Antony Rowe Ltd, Chippenham and Eastbourne

Contents

Notes on Contributors vii

Acknowledgements ix

Foreword x
Miriam Meyerhoff

Introduction: The Meeting of Gender, Language and Religion 1
Allyson Jule

Part I Gender, Language Patterns and Religious Thought **7**

1 An Overview of God and Gender in Religion 9
Münevver Tekcan

2 The Gender of God: Judeo-Christian Feminist Debates 25
Francis Britto

3 Asymmetries of Male/Female Representation in Arabic 41
Samira Farwaneh

4 American Women: Their Cursing Habits and Religiosity 63
Timothy Jay

Part II Gender and Language Use in Religious Communities **85**

5 Women and Men: Languages and Religion in Taiwan 87
Chao-Chih Liao

6 Women's Letters to the Editor: Talking Religion in a Saudi Arabian English Newspaper 101
Hannes Kniffka

7 A Cyber-Parish: Gendered Identity Construction in an On-Line Episcopal Community 133
Sage Graham

8 Language Use and Silence as Morality: Teaching and Lecturing at an Evangelical Theology College 151
Allyson Jule

9 The Children of God Who Wouldn't, but Had To 168
Annabelle Mooney

Part III Gender and Language Use in Religious Identity 185

10 'Restoring the Broken Image': The Language of Gender and Sexuality in an Ex-Gay Ministry 187
Amy Peebles

11 '*Assalam u Alaikum*. Brother I have a Right to My Opinion on This': British Islamic Women Assert Their Positions in Virtual Space 203
Fazila Bhimji

12 'Inshallah, today there will be work': Senegalese Women Entrepreneurs Constructing Identities through Language Use and Islamic Practice 221
Shartriya Collier

13 Gender, Hebrew Language Acquisition and Religious Values in Jewish High Schools in North America 240
Debra Cohen and Nancy Berkowitz

14 Speaking Our Gendered Selves: Hinduism and the Indian Woman 257
Kalyani Shabadi

Index 270

Notes on Contributors

Nancy Berkowitz, PhD, is an independent consultant in research, measurement and evaluation in Massachusetts, USA. She assists individuals, corporations and non-profit organizations with research design and data analysis, test development and programme evaluation.

Fazila Bhimji, PhD, is Lecturer in Humanities at the University of Central Lancashire, UK. Her research interests are multifold. She is currently working on gender, language, power and spiritual identities among Muslim women in Britain today.

Francis Britto, PhD, is Professor of Linguistics at Sophia University in Tokyo, Japan. His work is related to computer literacy, feminism, India, sociolinguistics, the Internet and religions. He has most recently served as Programme Chair for the JALT International Conferences.

Debra Cohen is a doctoral candidate at the Hebrew University, Jerusalem, Israel, examining the role of attitude and motivation among teenagers learning Hebrew. She has worked as a child therapist and educational psychologist in the United States and currently in Israel.

Shartriya Collier is a doctoral candidate in Language Education at Temple University in Philadelphia, USA, where she also lectures in bilingual and multicultural education. Her research interests include immigrant women entrepreneurs and language acquisition.

Samira Farwaneh, PhD, is Assistant Professor of Arabic Language and Linguistics in the Department of Near Eastern Studies at the University of Arizona, Tucson, USA. Her research interests include Arabic phonology, morphology and sociolinguistics, particularly language and gender issues.

Sage Graham, PhD, is Assistant Professor of English and Linguistics at the University of Memphis in Tennessee, USA. Her current research addresses identity formation and conflict in computer-mediated communication, the impact of technology on teaching effectiveness and medical discourse.

Timothy Jay, PhD, is Professor of Psychology at Massachusetts College of Liberal Arts. He has written several books and articles on cursing and psycholinguistics. His research deals with verbal aggression and the use of taboo language.

ALLYSON JULE, PhD, is Senior Lecturer in Education at the University of Glamorgan, Wales, UK. She has published several articles and a book on ethnic-minority girls in classrooms and is particularly focused on the use of gendered linguistic space among classroom participants.

HANNES KNIFFKA, DPhil, is Professor of General and Applied Linguistics at Bonn University, Germany. He has published several books and articles on sociolinguistics, 'culture-contrastive' and anthropological linguistics, textlinguistics and forensic linguistics.

CHOA-CHIH LIAO, PhD, is Associate Professor of English and Linguistics at National Chiayi University, Taiwan. Her research interests include cross-cultural communication, humorology, onomastics, as well as language and religion related to Chinese or Taiwanese societies.

MIRIAM MEYERHOFF, PhD, is Reader in Theoretical and Applied Linguistics at the University of Edinburgh. She is the author of several books and articles on language use and is an active scholar in the field of feminist linguistics.

ANNABELLE MOONEY, PhD, is a Research Associate at Cardiff University in the Centre for Language and Communication Research, Wales, UK. Her research interests include the rhetoric of marginal religious movements.

AMY PEEBLES, PhD, is Assistant Professor of Linguistics at Truman State University in Kirksville, Missouri, USA. Her research interests include language and identity, language ideology, gender, sexuality, religion and life history narratives.

KALYANI SHABADI, PhD, is a Researcher at the Resource Centre for Indian Language Technology Solutions, Indian Institute of Science, Bangalore, India. Her main areas of research are formal syntax, semantics, computational morphology and sociolinguistics.

MÜNEVVER TEKCAN, PhD, is Assistant Professor in Turkology at Kocaeli University in Turkey. Her current research activities include linguistic analyses of Middle Turkish and Chagatay manuscripts. She has a special interest in Central Asian Sufi texts.

Acknowledgements

The contributors in this collection are to be recognized for their massive efforts, particularly under tight summer deadlines, as well as for their exceptional and innovative research connecting gender, religion and language. It has been a great joy to work with each of them and to have such an impressive and varied collection of research projects all in one place. I particularly wish to thank Miriam Meyerhoff at the University of Edinburgh for her generous support and for her insightful and important comments in the book's foreword.

I would also like to thank the many people who helped in developing the concept for this book. The encouragement received by scholars at IGALA3 (International Gender and Language Association at Cornell University, June 2004) greatly helped to strengthen my reserves to bring this collection together. I am especially grateful to Jill Lake at Palgrave who encouraged the book, recognizing a vacuum in linguistic study connecting religion and discourse. Thank you too to Nikki Niles for her thoughtful cover photograph, and to Cheryl Wall and Penny Simmons for their help in the editing process. And finally, all of the contributors in this collection wish to extend their personal acknowledgements to the family and friends who continually support their work.

Foreword

Most people in the world today claim to be the follower of an organized religion. If anything, the proportion is increasing: in the last decade, the number of people who claim to be non-religious or atheist has dropped from approximately 1.1 billion to about 850 million. Most of the people who at least minimally identify with a religion are associated with one of the major faiths – Judaism, Christianity, Islam, Hinduism, Buddhism and Sikhism. Although there are internal differences within each of these families of faiths, these statistics indicate a remarkable degree of global conformity and they point to the potential for the development of enormously powerful supra-local identities. Religious identification is increasingly an issue in both international and local politics. In some traditionally secular societies, strong associations between individuals' religious beliefs and their position on non-religious issues such as capitalist individualism, reproductive rights and educational opportunity are creating a *de facto* erasure of traditional boundaries between church and state. In other communities, religion has been bound closely with these concerns for a long time, and increased opportunities for bodily and virtual mobility present challenges to established relationships between religious and other social identities as well.

Gender and the Language of Religion represents a timely move to explore exactly how ideologies of gender and religion intersect. It also considers how individuals negotiate and enact the daily practices of their lives when religion as well as gender is among the identities that they are orienting to. The contributions to this volume are varied, but all offer something of merit to the reader. We are given a perspective on the ways in which students and professors in an evangelical Christian college enact and enforce gender roles in their pursuit of devotion. We also find out how immersion in traditional Hebrew schools has a different impact on the acquisition of Hebrew by male and female students. Three papers discuss liturgical references to women and men and consider the impact these have on the lived experience of the adherents of different religions. There are fascinating papers on the way religion is foregrounded or backgrounded in different forms of discourse – the Internet, newspapers, life history narratives and of course in ritual language.

Readers will find *Gender and the Language of Religion* provides a healthy blend of methodological approaches. Some of the articles work within the more experimental paradigm of social psychology some provide quite personal, in-depth studies of interactions in a hairdressing salon, the classroom, or they offer close critical analysis of the discourse used in 'ex-gay' narratives. Some draw more on the traditions of literary criticism, and some on applied sociolinguistics.

What unites the papers in the volume is a fascination with how language, gender and religion come together as part of the backdrop of day-to-day life. As Jule puts it in her introduction: 'Religious life seeks transcendence from the mundane, but ... it is the seemingly mundane that also creates ... religious identities.' The authors in this volume invite us to transcend the typically secular boundaries of language and gender research and to take steps towards new frontiers in how we understand what two pivotal identities are for many people today.

Miriam Meyerhoff
University of Edinburgh

Introduction: The Meeting of Gender, Language and Religion

Allyson Jule

Understanding the role of religions in the world is not at all a matter of reading about exotic people and places or coming to some conclusions regarding versions of truth as expressed in various communities. In today's global pluralism, almost any faith can be found anywhere, both as a presence and as an option of faith. Hinduism, Buddhism, and Islam (religions originating in the East) are found all over the West, and the various representations of the West's Judaism and Christianity are now well established throughout the world. A glance at any newspaper or any TV newscast reminds us of how current events are embedded in religious communities and how expressions of religion deeply affect the concepts of diversity and globalization – or our resistance to such concepts. No comprehension of world affairs or the larger human condition is possible today without some understanding of the role of religion and how religion influences human behaviour. Most of us in the course of our lives will come into contact with people from a wide range of religious experiences, while at the same time our own religious experiences will influence others and our views of the world and the people who surround us. This book searches for a range of language experiences within religious communities beginning with a recognition of liturgy on to the working out of individual identity within religious groups. The rhetoric, the method of meditation and religious education differ from group to group but, regardless of these differences, people are influenced by the ways their culture attempts to affirm human life and attempts to transcend it. The practices of religion (such as worship, rites of passsage, forms of devotion, group activities) constitute religious expression and such religious expressions are woven into the cultures in which we live and into the way we live within them.

One of the ways in which religion today is in a state of flux and transition involves gender roles, particularly the role of women and the growing, shifting awareness of femininity and masculinity. Issues such as veiled Islamic women or female ordination in Christianity are two examples of both religion and culture grappling with gender roles. Because of common patriarchal roots in the world religions, women's roles have historically been very limited while men's roles have often been more developed or highly specific. I hope that, through survey books such as this one, perceptions or assumptions can be deeply explored, allowing for fresh perspectives on the role religion plays in society.

In reflecting on the position of religion and religious communities alongside the current context of language and gender research, the contributors in this collection offer us a variety of views and experiences – each entering the discussion at various points of interest and expertise. The book explores the ways we live with religion as a cultural discourse and how gender, language and religion intersect in various yet shared ways around the world. How language and gender are made meaningful, how gender is interpreted and lived inside religious communities, and how religion and gender impact on identity are all themes explored here.

In keeping with the local nature of each chapter, attempts have not been made to make the book more consistent concerning terms or their spellings (example: Koran or Qu'ran). This is to allow each author to represent the ideas from within certain groups. It is also not the primary goal here to review the criticisms of the larger debates in language and gender research and scholarship; such discussions are well articulated elsewhere. Instead, I wish to bring focus to religion within sociolinguistic study, something not yet adequately grappled with inside sociolinguistics. I believe the variety of scholarship in this collection creates a vibrant offering to the field. It appears religion sits well in the fields of sociology, anthropology and theology, but it is almost unexplored within linguistic research. As such, this collection specifically explores the three themes of gender, language and religion simultaneously, allowing for a larger reflection on how these elements work alongside each other in both complementary and contradictory ways.

Why the connection

The history of religious traditions often seems divorced from the more private and lived experience of faith. The spiritual quest is often an interior, personal journey while religion seems preoccupied with liturgy and doctrine. However, religions have a life outside theology and, as such, can be a lens through which to understand something of society and how it is we live, and live together, in certain ways.

Religious people themselves have a varied reputation. For some, religion is perhaps an admirable organizer of life and how to live it. For others, it is seen and dismissed as something imaginative or unnecessary. Regardless of one's own religious views or one's personal sense of faith, religion has been a force in the world – offering solace and peace of mind to a few, charging some with grand or charitable acts, and influencing others to violence or cruelty. Religious groups may fight between and among themselves over versions of truth and tradition, and religious leaders of all faiths are sometimes consumed by worldly ambition or human frailty. However, each religion articulates the universal quest to find something sacred amidst the secular. In the West, there has been an attempt to separate religion from politics – liberating religion from the corruption of politics or liberating politics from the corruption of religion. However, religious people around the world sometimes believe they have a duty to bring their ideals to bear on society or they are goaded to interact with society in an attempt to save it in some way because of the power religion has had in their own lives. As a result, religion becomes personal and a way to self-definition. And because we live together in groups, we are influenced by the religious views of those around us, in the midst of our own personal views or regardless of them.

Religious life seeks transcendence from the mundane and yet, as this book suggests, it is the seemingly mundane day-to-day living that also creates religion, creates religious communities, and creates religious identities. That religion is so intricately connected with culture or ethnicity is inescapable. There are converts who challenge this fact but, more often than not, one's religion and one's religious views grow out of one's culture of origin. We are religious creatures because we are compelled to find meaning in our lives. In doing so, we emerge with our own particular realities and seek out meaning for them in something larger, each in our own place and each inside our own frames of reference.

Gender, language patterns and religious thought

The four chapters in Part I introduce the reader to some foundational issues, including the authoritative language of religion and ways the holy texts have been understood or misunderstood concerning gender. Münevver Tekcan's chapter sets the stage by providing an overview of many of the world's religions – Judaism, Christianity, Islam, Hinduism, Buddhism, as well as Ancient Greek and Roman mythologies – and exploring the view of gender and the divine, each within each. Tekcan also discusses the particular cultural influences within each expression, suggesting that the various interpretations of 'God' have been influenced by the various interpretations of ourselves. Francis Britto's chapter, 'The Gender of God: Judeo-Christian Feminist Debates' offers an overview and critique of how God has been understood as male in Judaism and Christianity. His work explores the particular issues of concern within Christianity and the implications of seeing God as male or female. Both of these chapters explore the traditions of religion and how a language for gender has been understood and influencial. These first two chapters set up a starting point to our sociolinguistic look at religion language and gender.

The next two chapters reflect on the ways language is used alongside religion – more specifically, how language patterns reveal religious influences. Samira Farwaneh's work looks at the asymmetries of sexist language in Arabic and how modern media have influenced such patterns within Islam. Timothy Jay's work looks at how cursing, specifically 'Oh my God', is now part of women's speech patterns in the United States, more so than it is among men. Both Farwaneh and Jay give some reflection on how language use is connected with religious views and how language serves as part of an assertion of power and identity within them.

Gender and language use in religious communities

The chapters of Part II all examine gender and language inside religious groups or communities. Each brings sociolinguistic research to bear on the larger discussion of gender, language and religion. The chapters each explore various ways devout behaviour is gendered and how language provides some evidence that it is so. The first of the chapters is Chao-Chih Liao's work on women and men in Taiwan and how it is that Christianity and Buddhism attract followers because of the languages used – English as a way to both education and conversion. Hannes Kniffka explores how women in Saudi Arabia interact through Letters to the (male) Editor to articulate their religious experiences, and how the male editor constructs them as women in print. Sage Graham's work on women in an Episcopal church in the United States explores how their on-line discussions reveal and restrict their roles in church life. My own research inside an evangelical college in Canada explores how morality is gendered so that both men and women are rehearsed and rewarded into specific ways of being understood as moral: men as public speakers and women as polite audience members. Annabelle Mooney's work explores life inside 'The Children of God' cult-like community and how group membership is negotiated and explained from the inside.

All the chapters in Part II seem to agree that, regardless of the interpretations of theology or sacred texts explored earlier in Part I, many religious groups are in tension and in negotiation regarding gender and gender roles. There appears to be a metanarrative at work which serves to essentialize gender so that women are designated as devout and loyal, as seen in Liao's work in Taiwan and Graham's work in the American on-line parish; as quiet, as discussed in my own work inside a Canadian theology college; and as frustrated within religious groups, as expressed to some extent in Kniffka's exploration of Saudi women's letters to the editor and Mooney's work on the American group, 'The Children of God'.

Gender and language use in religious identity

Part III is a collection of chapters, each exploring the role of religion in creating and influencing individual identity. Amy Peeble's innovative work on an ex-gay 'therapy' group in Texas sheds light on how gender roles and sexuality roles are understood and then prescribed to community members. Fazila Bhimji looks at young Muslim women in Britain and how they negotiate themselves between and within multiple identities; Shartriya Collier looks at Senegalese Muslim women working in the United States; Debra Cohen and Nancy Berkowitz explore young Jewish teenagers, both male and female, learning Hebrew; and, finally, Kalyani Shabadi completes the collection with her look at Hindu women in India and how it is that they identify themselves as both Hindu and female.

This book

Religions share the ideas of icons, symbols, sacrifice, behaviour, attitudes and quest as part of a meaningful life. However, how we each explore and how we each relate to religion is infinitely individual, shifting from various places and times, and most times significantly embedded in culture and in communities.

In part then, this volume is an attempt to turn some attention within linguistics to the impact of religion as well as to perhaps turn some attention within religious studies to the impact of language use. In either case, gender is seen as a significant variable influencing both and, in turn, as being influenced by both. In all this, a new way of viewing linguistics is the point. Religious life and the assumptions around it and in relation to it are constructed in various ways and therefore need continual and ever-new ways of understanding. There is more or less general agreement among the contributors of this volume that gender, language and religion cannot be discussed separately. Because of the enormous influences of all three on all three, they can and must be discussed and explored together.

We live in a world where religion plays an enormous role in influencing a wide range of spheres, including community life. The various religions used for discussion here allow for an exploration of the various ways religion is interpreted and experienced through gender and through language use. The fact remains that social science must take into account the effects *of* language and the effects *on* language. In particular, one cannot understand sociolinguistics without some reflection on variables which affect language. Gender and religion are two; there are clearly other variables. Nevertheless, this collection provides some reflection on the effect these two enigmatic variables have on language use and the complex relationship gender and language have with religion.

Part I

Gender, Language Patterns and Religious Thought

1

An Overview of God and Gender in Religion

Münevver Tekcan

There are many different languages and ways to communicate and, similarly, there are many different religious concepts that have changed throughout history. On this we are agreed. However, there are significant concepts within the world religions where there is much uniqueness. One significant concept is that concerning the gender of God. Not all religions have a godhead; some have none while others have many. This chapter explores how various religions view the gender of the divine and how their languages evoke a common meaning.

For monotheistic religions, the deity is usually referred to in the masculine though often argued to represent both genders or to have no gender at all. Polytheistic religions have deities of both sexes and therefore easily represent both male and female. Meditative beliefs, such as Buddhism, concentrate on the individual's own physical and mental actions. Each religion also has its own language of teaching. Inviolable religious texts are paramount to religions such as Islam; in ancient Greek, Roman and Egyptian texts, gods and the divine take the form of characters in literature and in stories. Visual representations of gender range greatly, from erotic tantric Hindu sculptures to consistent Islamic calligraphy. As such, representations of gender and gender connotations concerning God/gods are complex and often so deeply experienced as to be subliminal or even subconscious.

Metaphor and irony are also available within religious literature to communicate religious thought. Visual language offers a wide range of alternative methods of representation and interpretation. All of these are subject to cultural influences. Within this complex and dynamic domain, there are five possible concepts of gender for a godhead:

1. 'God' has no gender
2. 'God' transcends gender
3. 'God' is both masculine and feminine
4. 'God' is masculine
5. 'God' is feminine

To Buddhists, the gender or even the existence of God is not as important as an individual's actions trying to reach *Nirvana*, an enlightened state where one is freed from greed, hatred and ignorance. To Muslims, God is transcendent, *tanzih*, over gender; Allah is beyond anything we know or can know (Murad, 1999). There is a common notion that God is unknowable and is beyond reach. Islam speaks of Allah in terms of deity without gender – above our comprehension and experience ('Concept of God in Islam', on-line). In the Koran, God is referred to as *Huwa*. Arabic has no neuter form so no gender is necessarily implied when Allah is referred to as He, any more than femininity may be implied by the grammatically female gender of other neuter plurals.

The word for 'spirit' is feminine in Hebrew, neuter in Greek and masculine in Latin. Language genders God in all three ways. According to Christianity, humanity was created in God's image, hence implying that God is either male or female, though it has been argued that the female aspects of the Judeo-Christian God have been overlooked. In order to femininize God, 'God is ... envisioned in roles taken from ... female experience such as midwife, nurse, seamstress, mistress of a household, and owner of money who searches for a lost coin that is very important to her, rejoicing with neighbours when it is found' (Johnson, 1984, p. 10). Yet as God is given female attributes, God may be lessened – endorsing woman's often lower status in society. What the linguistic symbols signify is governed by the conventions used to receive the message. Each religion has its own conventions, as does each culture and each part of society within various cultures.

Islam offers a similar image which describes how, during the Muslim conquest of Mecca, a woman was running about in the hot sun, searching for her child. She found him, and clutched him to her breast, saying, 'My son, my son!' The Prophet's Companions saw this and wept. The Prophet was delighted to see their *rahma*, (their emotion) and said, 'Do you wonder at this woman's *rahma* for her child? By Him in Whose hand is my soul, on the Day of Judgement, God shall show more *rahma* towards His believing servant than this woman has shown to her son' (in Murad, 1999, on-line). This image is seen to give Allah a female characteristic – emotion. Traditional male and female attributes are so entrenched in human culture and in language that it is easy to associate *rahma* with the word *rahim*, which means womb.

The male Hindu gods have female consorts. The Hindu and Buddhist tantric rituals are usually set as a dialogue between Lord Shiva and his consort Parvati. He explains to her the philosophy and myths underlying the tantric ritual. Tantric ritual involves reversals of the more common Hindu social practices; it reverses physiological processes, such as the drawing up of the semen out of the woman and into the body of the man rather than from the man into the woman. The female consort is seen as the principal force controlling the strength of the male. Within this Tantric tradition, there is a strong binary where the male is in the role of 'strength' and the female plays that of 'control'. Together, there is complete creation. In Hinduism, 'god' is not whole without both maleness and femaleness together.

In Christianity, 'God the Father' has been a dominant concept, while in Hinduism the many gods are male with female consorts. The masculine has associations of strength, power and dominance. A male god remains a male god even when displaying traits such as sensitivity, love and nurture. The Western concept of 'new man' contains masculine and feminine associations and yet remains 'man'. When trying to define 'new god', 'the introduction of presumably feminine features, the andocentric pattern holds. Since God is still envisioned in the image of the ruling man only now possessing milder characteristics, the feminine is incorporated in a subordinate way into an overall symbol that remains masculine ... the feminine is there for the enhancement of the male ... adding "feminine" traits to the male-imaged God furthers the subordination of women by making the patriarchal symbol less threatening, more attractive' (Johnson, 1997).

Egypt, Greece and Rome all had female gods that were equal to the males and held prominent positions. Isis was the Egyptian goddess of motherhood and fertility. A cult grew up around her and spread out from Alexandria in the fourteenth century BC to the Hellenistic world. This cult appeared in Greece in combination with the cults of Horus, her son, and Serapis (the Greek name for Osiris). The tripartite cult of Isis, Horus and Serapis was later introduced (around 80 BC) into Rome in the consulship of Lucius Cornelius Sulla and became one of the most popular branches of Roman religion.

Hinduism has many goddesses which are worshipped in their own right or alongside their male counterparts. The female goddesses are connected with letters and language. Within the Tantric tradition, where the female is the controlling force, religious texts use 'a type of language which can be taken on many levels ... everything has a gross, a subtle and a supreme meaning ... which can be taken at face value but do not always have this meaning' (Shiva Shakti Mandalam, on-line). Derrida (1976) believed that, after deconstruction, the real meaning of a text was in the realization of all the possible readings. Hindu texts seem to operate in a similar way; where the supreme meaning is dependent on its sub-meanings – something arrived at through the realization of all possible readings.

Different cultures influence gender experiences in different ways. Different religions have different relationships to gender. Within each religion, both genders have a myriad of issues relating to the God figure. Each culture and religion has its own language. Religious thought, teachings and concepts of the deity are communicated verbally, visually and through the various sacred texts. Religious groups have frequently disagreed with each other, and the treatment of gender and the deity is no exception. Several world religions will now be discussed in turn.

Christianity

Christianity has been using patriarchal language for centuries. On the surface, language, pictorial representations and iconography have been male dominated. Arguments have been put forward suggesting that 'the fatherhood of God is and must remain the predominant Christian symbol; it is not a closed or exclusive symbol but is open to its own correction, enrichment, and completion from other symbols such as mother' (Hooft, 1982, p. 133). Elizabeth Johnson (1997) sees this as improving the father symbol, but there is no female equivalent. She says 'the feminine is there for the enhancement of the male, but not vice-versa: there is no mutual gain' (p. 10). It is only recently that the Church has allowed women into the senior ranks but language has changed very little. 'The new priestesses have to deconstruct the old hierarchy and system of communication if they want a more holistic gendered God, though to some "deconstruction within theology writes the epitaph for the dead God" ' (Raschke, 1982, p. 27).

Parables or stories have played an important part of communicating the Christian message. Many religious leaders have used narrative to communicate complicated, philosophical truths. The message and meaning can be very different. The narrative style of a parable does not always carry a clear message. Isaiah 6: 9–12 says, 'Hear ye indeed, but understand not; and see ye indeed, but perceive not. Make the heart of this people fat, and make their ears heavy, and shut their eyes; lest they see with their eyes, and hear with their ears, and understand with their heart, and convert, and be healed.' The metaphors presented here make the text a complicated one.

Metaphors have a common meaning in the context of shared social experience, and it is questionable if male and female experiences are all that similar. Metaphors borrowed from sexuality and gender are used to describe the relationship between Father, Son and Holy Spirit: the metaphors involve procreation by and from God. Procreation arises from the ancient theory of procreation, according to which man begets life from his 'seed', while the woman is viewed as the 'receptacle', providing the shapeless 'material' for the new life (Raming, 1999, on-line). In the ancient world, the male was dominant; therefore, Christian story-telling has come from a dominantly male voice using male eyes in a man's world.

Early Christian art tried to communicate religious teaching to a largely illiterate population. These pictures tended to be simple and naive. They used a simple visual language illustrating events from the Bible. During the Byzantine Empire icons encrusted with gold and lapis lazuli portrayed the divine nature of religious figures. In the fourteenth century artists such as Giotto painted with yet greater depth of feeling.

Up to the Renaissance, artists only had a limited visual vocabulary. However, as art became more sophisticated, images used many techniques to display that the artist (and the patron) were in control (Gombrich, 1971). Caravaggio used theoretical chiaroscuro that went beyond reality. What Caravaggio was accused of, however, was having promoted Beauty and Truth while academic tradition sacrificed Truth for Beauty (Gombrich, 1971). Gendered semiotics began to play a greater role in Christian art, and works such as Bernini's *Ecstasy of Saint Teresa* demonstrate how, during the High Renaissance, the female image could be manipulated in a religious context. Production of religious imagery of the divine was the domain of the wealthy elite who alone could dictate content and style (McFague, 1987). Knowledge, learning and power had replaced gold and other precious materials as symbols of status. This power was in the hands of those born male.

A common form of Christian prayer begins 'Our Father, who art in heaven'. As a form of communal worship it emphasizes the male at the head of the family. Some see Mary, The Virgin Mother of Christ, as a restoration of the element of femininity to the godhead (Cope, 1959). 'Virgin' and 'mother' could be seen as a difference, where the meaning of God shifts on a feminine binary – a move towards a female godhead. However, The Virgin Mary is removed from one of the most basic female experiences by having a virginal conception. It has been argued that Christian texts not only reflect society at the time of Christ, but perpetuate male centred experience, ignoring and even dismissing the female experience (Børresen, 1995).

Even the first Judeo-Christian story of Adam and Eve is imbued with masculine centrality. Eve tempts Adam, and Adam blames Eve for the Fall of Man. It can be argued that any text and the language it uses embody the characteristics of the society it was produced in. Yet many religions tend to empower archaic texts with a special significance that in some way is related to the deity and is eternal.

Islam

The God in Islam, Allah, is said to be above gender. The deity is not male, female or even neuter. The deity is not androgynous or even without sex. Another way to describe gender and Allah is in terms of the Zen Buddhist one-handed clap. Arabic has no word for neuter or the sound of a one-handed clap. There is no dualism because 'He is God the One God, the Everlasting Refuge, who has not begotten, nor has been begotten, and equal to Him is not anyone' (Sura, 112 in Murad, 1999, on-line). Islamic teaching is based on the Koran which seems to be, and is viewed by Muslims as, the direct word of God. The Koran is the recitation of the word of Allah, a deity without identification of gender (Murad, 1999).

Oral transmission of religious thought is important to many religions but particularly Islam. This *phonocentrism* could be used to argue that the Koran is the closest one can get to the mind of The Divine. Structuralism suggests that the meaning of words is dependent on their difference to other words. 'Black' has meaning in its relationship with 'white', and 'male' in its relationship with 'female'. Allah and the Christian godhead stress their originality by emphasizing that The Divine is the creator, not the created. As such, there is nothing to complement The Divine in a binary. The Christian 'God the Father' is without a consort, and Allah stands isolated. However, Murad (1999) highlights, 'Izutsu and Murata, who have both noted the parallels between Sufism's dynamic cosmology and the Taoist world view: each sees existence as a dynamic interplay of opposites, which ultimately resolve to the One' (on-line).

It could be argued that there is pluralism in Allah, as seen in the 99 names given to Allah. Sufi metaphysicians have gendered names for

Table 1.1 Gendered names for different aspects of Allah in Islam

♂ **Names of Majesty (*jalal*)**	♀ **Names of Beauty (*jamal*)**
The Powerful (*al-Qawi*)	The All-Compassionate (*al-Rahman*)
The Overwhelming (*al-Jabbar*)	The Mild (*al-Halim*)
The Judge (*al-Hakam*)	The Loving-kind (*al-Wadud*)
The Absolute Ruler (*al-Malik*)	The Source of Peace (*as-Salam*)
The Victorious (*al-'Aziz*)	The Shaper of Beauty (*al-Musawwir*)
The Greatest (*al-Mutakabbir*)	The Humiliator (*al-Mudhill*)
The Creator of The Harmful (*ad-Darr*)	The Preventer of Harm (*al-Mani'*)
The Taker of Life (*al-Mumit*)	The Giver of Life (*al-Muhyi*)

different aspects of God which were split into two groups (Table 1.1). Considering the names of Allah, the gendering falls in line with conventional, stereotyped notions of male and female. Power and control are seen as masculine and compassion and love are seen as feminine. In this sense, Allah has both male and female characteristics.

Islamic art does not allow the representation of living creatures; instead, it has concentrated on geometric designs and calligraphy. Such forms of art do not depict the human form and so aspects of gender are non-existent in Islamic art. Calligraphy places stress on the written word and on literacy. Beautifully written Arabic scripts of the Koran can be considered as a form of art. Religious buildings are frequently decorated with calligraphy, thus the worshipper is surrounded and totally absorbed in a formal religious context. But even among those who do not read the calligraphic inscriptions on various materials, the writing serves as a type of picture, and the illiterate population (or those who don't know Arabic) can appreciate its artistic beauty, even without knowing what is said.

Hinduism

Hinduism has one Supreme God who has aspects represented by 33 *devas* also referred to as gods. Hindu gods are divided into two different groups; Brahmanic and Vedic. Each of the 33 *devas* is an aspect of a supreme god. Each aspect has a specific power and function in supporting the world. Hindu gods take different forms and are considered personal to the believer, emphasizing one's moods, feelings, emotions and social background. The concept of god is both formless and has many forms. According to various Tantric texts, there are 33 million aspects of *Devi* or Hindu gods. The most important are listed in Table 1.2.

Table 1.2 Aspects of Hindu gods

Brahmanic Gods (trinity)					
♂ Brahma	creator	♂ Vishnu	preserver	♂ Shiva	giver
Brahmanic gods					
♀ Saraswati (consort to Brahma)	speech, wisdom, learning	♀ Lakshmi (consort to Vishnu)	prosperity, purity, generosity	♀ Durga (consort to Shiva)	beyond reach, destruction
♂ Hanuman (monkey god)	courage, hope, intellect	♂ Krishna	destroyer of evil	♂ Ganesh[a] (son of Shiva + Durga)	knowledge
Vedic gods					
♂ Dyaush	bright sky	♂ Varuna	water	♂ Indra	thunder bolts
♂ Surya	sun, light	♂ Satitar	sun	♂ Soma	speech
♂ Agni	fire	♂ Vayu	air and wind	♀ Ushas	wisdom, dawn
♀ Prithvi	earth	♂ Parjanya	rain	♂ Varuna	sky, water
♂ Yama	death				

Source: 'The Goddess in the Tantrik Tradition', on-line.

Hindu gods do not have wives but do have consorts. The consorts are not separate from them, but they are considered different forms of being. 'They exist in perfectly evolved soul bodies, bodies which are not properly differentiated by sex. ... They are neither male nor female. To better understand these Divine Gods, we sometimes conceive them as being the man if they are strong in expression or the woman if they are gentle and compassionate' ('Do God and gods have gender', article taken from *Hinduism Today*, on-line). This repeats the human propensity to anthropomorphize. God may be above sex and above gender but in the process of humanizing, gender stereotypes are introduced to make the divine accessible. However, in Hinduism, aspects of gender are not rigid and a single god can have male or female halves.

One early tale in Hindusim goes like this: A sage, one of the ardent devotees of Lord Shiva, used to worship only Lord Shiva and not his consort, Shakti (Parvati or Durga). The goddess Shakti, being the goddess of destruction, pulled out the energy from the sage's body. He was unable to stand. He pleaded to Lord Shiva who provided him with a stick. On its support, he stood and still worshipped Lord Shiva alone. Shakti observed ritual austerity, pleasing Lord Shiva who granted her the privilege of being part of his form. So the Lord Shiva now appears male on the right side and female on the left side (ardhanArIshvarar, on-line).

This story operates on two levels. At the human level, humanity is powerless compared to the gods. A god takes away power or gives it. In spite of all hardships, human beings must not stop their devotion, even though rewards may not be seen in this life. Shakti takes away the sage's power, yet she becomes part of Lord Shiva who seems to need and desire Shakti for himself to become complete. Together, they are seen as one individual. The gods act both as signifiers and as a metaphor for completeness. Male gods have powerful female consorts 'but the symbolic ascendancy of the feminine often goes with social denigration and low status of women in everyday life' (King, 1995, p. 16). This is not the case in Lord Shiva and Shakti.

The main religious texts for Hinduism are the Vedas – the oldest consists of 1028 hymns to a pantheon of gods. It has been memorized syllable by syllable and preserved orally. The two main gods, Shiva and Vishnu, are both male, with clear male attributes, but there are also a number of important female consorts. The Vedas are regarded as revealed canon (*shruti*), and no syllable can be changed. According to some language theorists, oral language can be closer to the original thought than the written word may be able to express. For Hindus, there is no possibility of changing the words or making modern translations.

Shiva personifies both the destructive and the creative forces of the universe. As the destroyer, he is represented wearing a necklace of skulls and surrounded by demons. His reproductive aspect is symbolized by the lingam, a phallic emblem. Shiva is the god of asceticism and of art, especially dancing. Vishnu is the god whose navel is a lotus which gives birth to the creator (Brahma). Brahma is the preserver of the universe and is worshipped in the form of a number of incarnations. Several of these are animals: the fish, the tortoise and the boar. Others are the dwarf, Vamana; a man-lion; the Buddha; Rama-with-an-Axe, Parashurama, who beheaded his unchaste mother; and Kalki, the rider on the white horse who will come to destroy the universe.

The goddess Devi and her various aspects are the main female deities. Devi is the prime mover who commands the male gods to do the work of creation and destruction. Appearing as Durga, she is the Unapproachable and kills the buffalo demon, Mahisha, in a great battle. Appearing as Kalishe, she is black and dances in a mad frenzy on the corpses of those she has slain and eaten, adorned with the still-dripping skulls and severed hands of her victims. Both male and female aspects represent the forces of nature in Hinduism and, in this respect, there is a close tie with the planet. The images operate more on the subconscious level than as clear metaphor.

The final parts of the Vedas are called Upanishads, which are speculative and mystical scriptures. 'The life of a mythology derives from the vitality of its symbols as metaphors delivering not simply the idea, but a sense of actual participation in such a realisation of transcendence, infinity and abundance, as this of which the Upanishads authors tell. Indeed, the first and most important service of a mythology is this one, of opening the mind and heart to the utter wonder of all being' (Campbell, 1990, p. 7). Male and female are not stereotyped; they are power structures in their own right.

Two of the most popular books within the Vedas are the *Ramayana* and the *Mahabharata*. The *Mahabharata* has been adapted for a play in the West End of London and one of the religious texts has been turned into a 'soap opera' on Indian TV, where it was so popular that extra story-lines had to be made up. Such modern popularity in the West demonstrates that some Hindu texts can be and are adapted to different contexts and, at the same time, remain fixed in ancient beliefs. Religious language operating in a pluralistic domain offers many interpretations for gender, even within a tradition of constancy.

Buddhism

Early Buddhism developed as a movement within Hinduism. The Buddha rejected the Hindu claim that a person's worth was set at birth and, in response, accepted all castes into his teachings. According to the Buddha there are Four Noble Truths to life:

1. Life is suffering, in its very nature, human existence is essentially painful from the moment of birth to the moment of death leading to further rebirth.
2. All suffering is caused by ignorance of the nature of reality and the craving and attachment that result from such ignorance.
3. Suffering can be ended by overcoming ignorance and attachment.
4. The path to the suppression of suffering is the Noble Eightfold Path:
   - right views
   - right intention
   - right speech
   - right action
   - right livelihood
   - right effort
   - right-mindedness
   - right contemplation.

Human actions lead to rebirth; good deeds are inevitably rewarded and evil deeds punished. Thus, neither undeserved pleasure nor unwarranted suffering exists in the world, but rather there is a universal justice.

What is important regarding gender, God and the language of religion is that for Buddhists the godhead is not all that important, and so the deity's gender is also unimportant. There are no symbols of mythical or real gods. The main canon, the *Tripitaka*, concentrates on discussions and rules on how best to reach a *Nirvana*, a state of consciousness beyond definition. As such, the gender of the divine is irrelevant entirely.

Buddhism is different to most other religions because the godhead only plays a minor role; Buddhism places emphasis on transcending the cycle of birth, death and rebirth, called *samsara*, and, as such, gender differences and inequalities are something to be overcome. Yet this denies one of the most basic of human experiences. Wilson (1996) argues 'that Buddhists have gendered *samsara* as female, and entrapment in it as a male dilemma. Because women have wombs, they embody rebirth, or that which Buddhism aims to transcend' (p. 71). Wilson seems to see a *femme fatale* in the woman's position within Buddhism and highlights how entrenched gender is in human culture.

Ancient Greece, Rome and Egypt

Ancient Greece and Rome have mythologies that still hold an important place in contemporary culture. The stories have a plethora of gods with heightened human values. The gods grapple with ethical and moral questions. Claude Levi-Strauss, the structural anthropologist, saw the gods and mortals in mythology acting as symbols 'to make sense of the world and to resolve cultural dilemmas' (*What are Myths?*, on-line). As in present-day 'magical realism', the incredible becomes more credible because it is so incredible. Once the incredible is accepted as that, the normal expectations can be broken. It is then possible to believe the extraordinary (Flores, on-line).

Drama was a recognized art form to the ancient Greeks and Romans – a prime way to explain the governing myths. Homer's *Iliad* and *Odyssey* are centred on the Trojan War; Euripides wrote a play centred on Helen. Helen was the most beautiful and intelligent woman alive and daughter of Zeus and Leda. The three goddesses Hera, Athena and Aphrodite asked the Trojan prince, Paris, to choose the most beautiful among them. Paris awarded the golden apple to Aphrodite, who had promised him Helen. Aphrodite persuaded Helen to leave her husband Menelaus, King of Sparta, and elope with Paris to Troy. Nine years of war followed. Helen was called to watch the final battle between Paris and Menelaus. Aphrodite helped Paris escape and, enveloping Menelaus in a cloud, she took him safely back to Helen.

In contrast to conventional religions, the Greek gods tend to serve as warnings that absolute power corrupts absolutely (Burkent, 1985). Human war is started by immoral gods. Helen, the most beautiful and intelligent woman alive, is corrupted by the goddess of love. Helen's beauty is traded for a golden apple; even nature is corrupted by money. The main protagonists in this story are female; the main victims are male – men who have to fight and die in the war. The final battle is prevented by the human values of Helen. Moral order collapses for the Greek gods as democracy emerges in Athens.

History, mythology and literature overlap in Greek and Roman mythology. Gods are used to make sense of the futile irony of war and dogmatic beliefs and to draw focus on society's attitude. Readings can take place on many levels. Euripides used everyday language, an antithesis of religious language which challenged dogma. In *Helen,* women take the dominant role. Euripides was writing at a time when beliefs such as that of the sun being a divinity began to make way for knowledge of the sun as made of matter. Myths, like religions in general, help explain the world.

Table 1.3 lists the characteristics and symbols associated with the 12 principal Gods in ancient Greece. Carl Jung's (in Campbell, 1990) notion of archetypes or primordial images can be seen in this list.

Table 1.3 Characteristics of ancient Greek gods

Twelve Principal Gods (Olympians)			
♂ Zeus (Jupiter)	sky, rain, cloud gatherer, eagle, oak	♀ Hera (Juno)	marriage, childbirth, protector of women
♂ Hephaisetos (Vulcanus)	fire, metalwork, craftsmen	♀ Athena (Minerva)	Greek cities, industry, arts and crafts, war and peace, wisdom, agriculture, owl
♂ Apollon	light, prophecy, medicine, music, poetry, pastoral arts, archer, athlete, agriculture	♀ Artemis (Diana)	chastity, marriage, childbirth, nature, harvest, wildlife, moon, youth
♂ Ares (Mars)	brutal nature of war	♀ Aphrodite (Venus)	love, beauty, fertility
♂ Mercurius	commerce, trade, athletes, good luck, wealth	♀ Histia (Vesta)	health, personal and communal security, happiness
♂ Poseidon (Neptune)	sea, powerful, vengeful	♀ Demeter (Ceres)	corn, harvest, fertility

Female gods represent childbirth, even though there is no particular reason for the female deities to represent marriage and fertility. The gender divisions between the gods tended to reflect societies' norms and the *status quo* power structure. The gods' immoral behaviour did not seem to impede their authority. 'Immortal, everlasting gods guarantee continuity, ritual means determination. Even the festival of dissolution and upheaval leads to the confirmation of the existing order ...' (Burkent, 1985, p. 21). Hera was able to avenge Zeus's numerous mistresses and offspring, and Athena is represented as armoured and wearing her breastplate, the aegis. Other female deities were also able to reap revenge on their unfaithful husbands. Male deities clearly held the power, but both genders held symbols of knowledge and art. Together with the narrative of mythology, human complexities are enacted out.

Linguists consider language as made up of signs that have meaning. In Egypt's Middle Kingdom (2134–1668 BCE) animals such as crocodiles, lions, sphinxes and bulls had connections with royalty which could be utilized in place of the literal word (Brass, 2004). In Egyptian mythology, each god is represented by the head of a bird or an animal.

The god Isis was the goddess of motherhood and fertility and the daughter of Keb (earth) and Nut (sky). She is frequently represented wearing the horns of a cow. Her powerful charms resurrected her brother-husband Osiris, who became the ruler of the underworld. Associations with fertility are almost universal and can be traced back to palaeolithic times where statues were made of pregnant-looking women.

The table of Egyptian mythology below (Table 1.4) shows that the female deities are strongly linked with fertility, nature and love. Several are represented by the head of a cow. Masculine deities are represented by the head of hunting birds. During the New Kingdom (1570–1070 BCE), 'the intellectual climate ... stressed a changing relationship between mankind and the gods. There is a unity of the cosmos where sub-ordination is transferred from individual-society to humankind-god and the protector is not a man but an omniscient deity who distributes favours and disgrace in reciprocity for the actions of individuals' (Brass, 2004, on-line). Religious texts, such as the *Book of the Heavenly Cow*, eventually gained more influence. This is regarded as the world changing from a primordial state where humans were detached from the gods to a state of imperfection but connection. Ancient mythology, though not a religion in the traditional sense, shows strong gender roles and assumptions. These roles and gender stereotyping have existed for many thousands of years, so it is not surprising that they still exist today in most modern religions. Language and symbols have changed, but an underlying

Table 1.4 Egyptian deities

Name	Gender	Attributes	Head	Hieroglyph
Ra (Re)	♂	sun	hawk	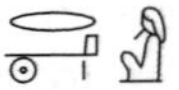
Osiris	♂	male productive force in nature, king of the land of the dead	man wrapped in a mummy's bindings	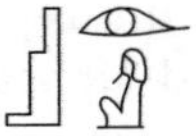
Isis	♀	fertility, motherhood	solar disk, cow horns	
Horus	♂	the day, sky, light, goodness	falcon	
Anubis	♂	embalming	jackal	
Hathor	♀	sky, heaven, love and gaiety, fertility, marriage	crow	
Shu	♂	atmosphere	human	
Seb (Geb)	♂	earth	goose	
Tefnut	♀	atmosphere, moisture	lioness	
Nut	♀	sky	naked woman	
Ammon	♂	reproductive forces	ram	
Mut	♀	mother, sky	vulture	
Thoth (Djeheuty)	♂	wisdom, writing, mediation	ibis	
Ptah	♂	creator, craftsmen	human	
Set	♂	darkness, evil	falcon	

'nature' remains constant. Gender is not a simple masculine and feminine issue; it has human complexities which are played out through time and place. The language of ancient cultures may be long forgotten but their symbols still have the power to communicate a sense of the world that continues to resonate as well as fascinate.

This chapter has examined the concept or nature of a god-figure. In spite of many differences between religions, there is a commonality of belief concerning gender and a desire to touch the divine, however it is constructed. Male and female have to communicate about their particular god-figure. They do so linguistically and symbolically in a context where language is influenced largely by culture. While deities and religions differ, humans are universally male and female and the experience of masculinity and femininity is common to all.

References

Adams, D. J. (1997). *Towards a theological understanding of postmodernism.* Metanoia (Prague), Spring–Summer. www.aril.org/adams.htm

Al Faruqi, L. (1999). *Islamic traditions and the feminist movement.* www.islam101.com/women/feminism.html

ardhanArIshvarar, siddhanta.shaivam.org/maardh.html

Barthes, R. (1970). *Mythologies.* Paris: Seuil.

Bass, C. (1978). *Writing and difference.* New York: Routledge & Kegan Paul.

Børresen, K. (1995). *Woman's studies of the Christian tradition: New perspectives.* In: U. King (Ed.), *Religion and gender.* Oxford: Blackwell (pp. 225–55).

Brass, M. (2004). *Remembering Egypt: Historical perspectives on the social construction of the image of Egypt.* www.antiquityofman.com/Egyptian_social_construction_of_image_of_ Egypt.pdf

Burkent, W. (1985). *Greek religion.* Cambridge, MA: Harvard University Press.

Campbell, J. (1990). *The inner reaches of outer space: Myth as metaphor and as religion.* In: New York: Harcourt Brace Jovanovich. *Concept of God in Islam.* www.ceet.niu.edu/ss/syed/concept.html

Cope, G. (1959). *Symbolism in the Bible and the church.* New York: SCM Press.

Culler, J. (1982). On deconstruction: Theory and criticism after structuralism. In: C. Broitman (Ed.), *Deconstruction and the Bible.* Ithaca, NY: Cornell University Press (pp. 93–4).

Deleuge, G. (1994). *Differences and repetition.* New York: The Athlone Press.

Derrida, J. (1976). *Of grammatology.* Trans. Spivak Gayatri Chakravorty. Baltimore, MD: Johns Hopkins University Press.

Do God and gods have gender? Article taken from *Hinduism Today,* www.hindunet.org/god/ht_article_god3.html

Flores, A. (1955/2003) Magical realism in Spanish-American fiction. In: L. P. Zamora and W. B. Faris (Eds), *What is magical realism?* (pp. 113–16). www.magicalrealism. com/what_is_magical_realism.php

Gombrich, E. H. (1971). *Norm and form: Studies in the art of the Renaissance* New York: Phaiden Paperback.

Greenstein, E. L. (1989). Deconstruction and Biblical narrative. In: C. Broitman (Ed.), *Deconstruction and the Bible.* Prooftexts 9 (p. 51).

Gunning, T. (2004). *Sacredness.* www.materdei.ie/logos/sacredness_article.htm

Hooft, W. A. V. (1982). *The fatherhood of God in an age of emancipation.* Geneva: World Council of Churches.

Johnson, E. (1984). *The incomprehensibility of God and the image of God male and female*. Theological Studies, vol. 45, no. 3, www.womenpriests.org/classic/johnson3.htm

Johnson, E. (1997). *She who is: The mystery of God in feminist theological discourse*. New York: Crossroad.

Johnson, P. A. and Kalven, J. (1988). With both eyes open: Seeing beyond gender. New York: Pilgrim Press. Reprinted/in: U. King (Ed.) (1995). *Introduction: Gender and the study of religion*. Oxford: Blackwell.

Joy, M. (1995). God and gender: Some reflections on women's invocations of the divine. In: U. King (Ed.), *Religion and gender*. Oxford: Blackwell (p. 129).

King, U. (Ed.) (1995). *Religion and gender*. Oxford: Blackwell.

Lambros, N. (2002). *God, truth and meaning in the postmodernism of Umberto Eco*. www.themodernword.com/eco/eco_papers_lambros.html

McFague, S. (1987). *Models of God: Theology for an ecological, nuclear age*. Philadelphia: Fortress Press.

Murad, A. H. (1999). *Islam, irigaray, and the retrieval of gender*. www.masud.co.uk/ISLAM/ahm/gender.htm

O'Connor, J. (1995). *The epistemological significance of feminist research in religion*. In: King (Ed.), *Religion and gender* (pp. 45–63).

Otto, R., *The idea of the holy*. In: *Rudolf Otto (1869–1937)*. homepages.which.net/~radical.faith/thought/otto.htm

Raming, I. (1999). Male discourse about God in the liturgy and its effects on women. In: *Lumen Vitae, Revue Internationale de Catéchèse et de Pastorale 55*. Trans. J. Waller (1999). www.womenpriests.org/theology/raming1.htm

Raschke, C. (1982). The deconstruction of God, deconstruction and theology. In: King (Ed.), *Religion and gender* (pp. 121–43).

Romans, S. (1994). *Language in society: An introduction to sociolinguisitcs*. Oxford: Oxford University Press.

Ross, K. L. (1996). *The Kant-Friesian theory of religion and religious value*. Paper delivered to 'The Roots of Rudolf Otto's Theory of Numinosity in Immanuel Kant, Jakob Fries, and Leonard Nelson', to the Philosophy of Religion section of The Southern California Philosophy Conference at the University of California, Irvine, 26 October 1996. www.friesian.com/numinos.htm

Saussure, F. de. (1916). *Course in general linguistics*. Trans. W. Baskin (1977). Glasgow: Fontana. courses.essex.ac.uk/lt/lt204/saussure.htm

Shiva Shakti Mandalam. www.shivashakti.com

The goddess in the Tantrik tradition. www.shivashakti.com/goddess.htm

The Islamic World to 1600. (1998). The University of Calgary: The Applied History Research Group.
www.ucalgary.ca/applied_history/tutor/islam/learning/calligraphy.html

Tillich P. (1957). *Dynamics of faith*. New York: Harper & Row at *What are myths?* www.livingmyths.com/What.htm

What are Myths? (2004). www.livingmyths.com/what.htm

Wilson, L. (1996). *Charming cadavers: Horrific figurations of the feminine in Indian Buddhist hagiographic literature*. Chicago: University of Chicago Press.

Young, K. K. (1987). Introduction. In: A. Sharma (Ed.), *Women in world religions*. Albany: State University of New York Press.

2

The Gender of God: Judeo-Christian Feminist Debates

Francis Britto

The cover of *Sex and God* (Hurcombe, 1987) features a provocative painting. It is the figure of a squatting human, having a flowing, triangular beard, a drooping belly, sagging breasts and four curvy hands. On this creature's head is another human face perched amidst matted locks of hair, and between its legs juts out yet another human face, surrounded by what could be taken as the main figure's pubic hair. An explanatory comment inside the cover reads that the figure visualizes 'the bodily birth (God as Mother of the entire fabric of creation) and the mental birth (God as the Father of the creative mind which must be "spoken" for ideas to be born)' (p. ii). Orthodox Christians may find the picture blasphemous and disturbing, but more and more women are being drawn to such images of God in place of the traditional grandfatherly image propagated by Judeo-Christian religions.[1]

What has caused this alienation of intellectual women from Churches? What is the problem with the gender of God, when almost all believers acknowledge that God is a pure Spirit beyond any gender? What consequences does this debate have on Churches and Christians? Such are the questions that this chapter aims to explore.

Feminist involvement in Church affairs

Without doubt, the gender of God has become a controversial issue because of the phenomenal success of feminist[2] movements. It is feminists who coined or redefined an artillery of words such as *sexism, androcentrism* and *patriarchy*, and argued that the contemporary social structure is sexist and patriarchal (e.g., Millet, 1970; Gornick and Moran, 1971;

Swindler, 1972; Emswiler and Emswiler, 1974; Lakoff, 1975; Ruth, 1980; Kramarae, 1981; Fiorenza, 1983; Frank and Anshen, 1983; Smith, 1985; Spender, 1985). Feminists were originally concerned with the progress of women in the secular world, but they gradually came to get involved in religion and confront the Churches because of various circumstances.

First, it has been customary for most authorities, who are predominantly male, to oppose feminists using the Bible as an authority. Such use of the Bible against women's franchise was what motivated Elizabeth Cady Stanton in the ninetheenth century to brand the Bible as the source of woman's subjugation and to create her own expurgated version (Hole and Levine, 1980; Fiorenza, 1992a). Some feminists even go so far as to suggest avoiding the Bible 'like the plague' or attaching to it the statutory warning, 'Caution! Could be dangerous to your health and survival' (Martin, 1991, p. 403). The questionable use of the Bible has plagued Christian Churches for several centuries and is current even today so that some feminists feel they have to combat what they perceive as Bible-based oppression.

Second, some feminists subscribe to what might be called a theory of 'male conspiracy,' which holds, with bits and pieces of historical evidence (Starhawk, 1979, 1992; Haddon, 1988; Morton, 1992; Stone, 1992), that human society was initially matriarchal and goddess-worshipping, and that, with the ascent of Judaism and subsequently of Christianity and Islam, matriarchy came to be destroyed methodically and deliberately by males. The whole of recorded history, therefore, is seen by these feminists as a history of female oppression and male conspiracy to keep women oppressed (discussed in Goldenberg, 1979; Spender, 1985; Walker, 1987). Since religion has played an important and authoritative role in recorded history, feminists hold religion chiefly responsible for women's inferior status. In the words of Collins, 'Discrimination against women and denial of power to them in the public political sphere has long been reinforced by organized religion' (1992, p. 149).

Third, the past few decades have witnessed an explosive growth in the number of highly educated women and women academics, especially in Europe and the United States. Women's achievements in the field of religion, despite the fact that they cannot occupy authoritative positions within the Church, have been simply astounding. According to one estimate:

> [B]etween 1976 and 1986, the total enrollment of women in programs leading to ordination (usually the M. Div.) rose 110 percent,

> whereas total seminary enrollment rose only 31 percent ... the number of women who have graduated ready for ordination has at least doubled and in some cases tripled, even within the Southern Baptist denomination, which traditionally does not ordain women. Overall, the number of women eligible for ordination increased 219 percent between 1976 and 1986, whereas the number of eligible men rose by only 7.8 percent ... In 1972, about 10 percent of the total student population were women. By 1986 the number of women studying theology had more than quadrupled.
>
> (LaCugna, 1992, p. 241)

Christian feminism evolved from the labours of such theologically educated women. One consequence of the increase in the number of women scholars has been the dispersion of the idea that the Bible, early Church fathers, leading theologians, popes and the Canon Law have all been contaminated by sexism and androcentrism (Daly, 1973, 1992; O'Faolain and Martines, 1973; Ruether, 1974, 1983, 1985; Fiorenza, 1983; Fiorenza and Collins, 1985; Martin, 1987; Vasquez, 1989; Ranke-Heinemann, 1990; Christ and Plaskow, 1992). The gender of God came to the fore in this context of general feminist disillusionment with Judeo-Christian religions. As they traced the roots of androcentrism, feminists perceived a relationship between the androcentric practices of the Churches and the androcentric nature of the Christian belief in God.

The maleness of the Judeo-Christian God

'Although Christians (and sometimes Jews) argue that God is without sex, neither female nor male, that contention is contradicted by a host of beliefs indicating the maleness of their God' (Ruth, 1980, p. 391). That the Judeo-Christian God is a male is, strangely, not a point that is vehemently denied, but enthusiastically defended by most traditionalists (see note 2). P. Mankowski, a Jesuit priest opposed to radical feminism, for example, says:

> The acknowledgement of God as Father is an essential part of Christian *kerygma* ['official proclamation']; it is unarguably the belief of the Catholic Church. The priest may responsibly take prudent measures not to give casual offense, but if he 'adapts' the wording to 'Parent' or 'Mother/Father,' he has forsaken that very doctrine which he was entrusted to pass on in the liturgy.
>
> (Quoted in Kaczor, 1992, p. 17)

G. A. Buckley, a Dominican priest, says likewise, 'God is preeminently both male and female ... Nevertheless, when God revealed himself throughout the history of the Old Testament, he revealed himself as male, as Father' (1991, p. 14).

What disturbs feminists is not that God is pre-eminently both male and female, but that the ecclesiastical language and practices fail to do justice to this belief. The traditionalists maintain that the use of masculine gender for God – in songs, prayers, liturgies and discourses – makes no assertion about the sex of God, but still its use is mandatory; whereas feminists maintain that precisely because God has no sex, the exclusive use of masculine gender is misleading and discriminatory, and so must be modified, complemented or given up.

That the Judeo-Christian tradition addresses God most often as a male can hardly be disputed. God in this tradition is usually given masculine titles such as Father, King, Lord, Bridegroom, and almost never feminine titles such as Mother, Goddess, Lady or Bride. Christian faith asserts that there are three persons in godhead, namely Father, Son and the Holy Spirit, and each of them is generally treated as a male – though the gender of the Holy Spirit is often debated (see Santini, 2001). The male metaphors are so pervasive that they can be found even among Christians of diverse languages and cultures. For example, even in a language like Japanese in which pronouns are rarely used, God is most often referred to as a male.

There are, surely, little-known passages in the Bible containing female metaphors for God (see Johnson 1989, 1992), but these have played only a negligible role in Christian history. It is, therefore, rare to find any Christian prayer, hymn, encyclical or theological treatise in which God is hailed as 'Mother' or 'Mother and Father' (for rare examples of recent popes referring to God as Mother, see Johnson, 1989, p. 520; Guimarães, 1999). The only female of some supernatural stature is the mother of Jesus, Mary, called since the Council of Ephesus in AD 431 *theotokos* 'Mother of God', but even Mary is not a goddess or a divine person of the Holy Trinity. As Epiphanius, bishop of Salamis, said in AD 403, 'The body of Mary is holy but she is not God ... Let no one adore Mary' (quoted in Johnson, 1989, p. 507). Although there have been Christian mystics such as Julian of Norwich who looked upon God and Jesus Christ as females (cf. O'Brien, 1964, p. 234), most Christians even today find it difficult to do so. As Lissner points out, 'To call God father and mother in songs, prayers, and liturgical texts will be difficult, if not impossible, for many believers for a long time' (1991, p. 314).

Traditionalist defence of male metaphors

A concise article that brings together the most common arguments for preserving exclusive male metaphors for God is Chris Kaczor's 'Inclusive language and revealed truth' in *Homiletic and Pastoral Review*, a conservative Catholic journal. According to Kaczor:

> The strongest reason to call God Father, from the Christian point of view, is that God himself through the Scriptures has told us how to speak of him. In all Scripture, in both New and Old Testaments, God is never called Mother. God is sometimes said to be like a mother, but Scripture never says that God is a mother.
>
> (1992, pp. 18–19)

Another argument of Kaczor is that Jesus called God only Father and so it is unwarranted to call God Mother. He cites from a book by William Oddie that '[i]n the whole of the Old Testament, God is described as "Father" only 11 times. Jesus, in startling contrast, uses the term at least 170 times, and, except for the cry of dereliction from the cross *always uses this form of address and no other*' [emphasis in the original]. This fact, concludes Kaczor, can only mean that 'calling God Father was an integral part of Christ's revelation' (p. 19).

E. A. Johnson (1992) points out several problems with this line of reasoning. Why, for instance, must one take the Father metaphor so absolutely when one ignores other serious metaphors that Christ used and other injunctions that Christ gave? The imminent Kingdom of God, for instance, was a major theme in Christ's teachings, but the Church does not present it today with equal urgency or vigour. The 'Father' image was not the only one that Jesus used to refer to God. His parables offer such a variety of images as the following for God: 'A woman searching for her lost money, a shepherd looking for his lost sheep, a baker woman kneading dough, a travelling businessman, the wind that blows where it wills, [and] the birth experience that delivers persons into new life' (p. 80). As to whether Jesus himself used the word *Father* as often as given in the gospels, there seems to be evidence that he did not. Johnson points out that the use of *Father* increases gradually in the Gospel narratives, leading scholars to believe that the more frequent use of Father in chronologically later texts reflects the theological developments in early Church rather than the frequency with which Jesus himself used the word.

Kaczor, however, also argues that the word *Mother* is etymologically incongruent with God:

> The science of etymology ... confirms that a Creator God cannot properly be called mother. Mother comes from the Latin word *mater*. From the Latin *mater*, comes *materia* rendered 'material' in English. All material is part of creation. Therefore, to call God Mother or *Mater* is to call the Creator the creation. This name change brings a slip from theism to pantheism, from Catholicism to New Ageism.
>
> (1992, p. 18)

Kaczor continues to say that religions which consider God Mother are all historically 'pantheistic' and 'paganistic'; they deny an absolute creation and an absolute Creator.

Kaczor's reasoning is etymologically interesting, but theologically unsatisfying. In Christian tradition, several metaphors (such as Rock) are used to describe God, but such usage has never been condemned as identifying The Creator with the created. Neither can calling God 'Father', 'King' and 'Master' then be acceptable since all these terms are applied pre-eminently to created human beings. Besides, Christians believe that God is everywhere, including in creation, and so one can argue that associating God with creation highlights this fact. Finally, some mystics, like Julian of Norwich, have called God 'Mother', and they are revered as great masters, not condemned as pantheists (cf. McLaughlin, 1992).

Feminist objections to male metaphors

In general, traditionalists defend the custom of treating God as a male with arguments based on doctrines, dogmas and scriptures. They believe strongly that the hierarchical Church is the guardian of an *immutable deposit* of faith, that the male metaphors for God sanctioned by Scripture and Tradition are part of this deposit, and that any attempt at modifying this deposit would be sacrilegious and heretical (Steichen, 1985, 1992; Ferreira, 1987; Kaufmann, 1992; Kelly, 1992; Ostling, 1992).

Feminists, on the other hand, challenge the male identity of God with arguments based rather on contemporary norms of justice and equality than on doctrines, dogmas or scriptures. 'Feminist method is based on a "hermeneutics of suspicion." Hermeneutics sets forth criteria for interpretation. A hermeneutics of suspicion regards all theological claims produced by the patriarchal mind-set to have an underlying bias in favor of men' (LaCugna, 1992, p. 243). Even moderate Christian feminists, therefore, would question the legitimacy of Scripture, Tradition and the Magisterium to dictate norms that keep women subservient. As proclaimed by Ruether, a feminist theologian, whatever 'denies, diminishes or distorts the full humanity of women is not redemptive. Whatever does promote the full humanity of women reflects true relation to the divine' (in LaCugna, p. 243).

Many books have been written by feminists regarding their reasons for why the male-only metaphors for God must be given up (Daly, 1973, 1992; Goldenberg, 1979; Halkes, 1981; Metz and Schillebeeckx, 1981; Johnson, 1984, 1992; Haddon, 1988). In essence, their reasoning parallels that of feminist linguists who argue against the generic use of masculine forms in language (Martyna, 1980; Spender, 1985). Like feminist linguists, feminist theologians contend that male metaphors for God reinforce androcentrism, subordinate women to men, make women invisible or unimportant, and silence them (Christ and Plaskow, 1992); as theologians, they also argue that male-only imagery is idolatrous and theologically untenable. The standard feminist position is concisely stated by Johnson:

> Patriarchal God symbolism functions to legitimate and reinforce patriarchal social structures in family, society, and Church. Language about the father in heaven who rules over the world justifies and even necessitates an order whereby the male religious leader rules over his flock, the civil ruler has domination over his subjects, the husband exercises headship over his wife.
>
> (1992, pp. 36–7)

Some of the most commonly cited instances of sexism in the Church and its authorities are:

1. The biblical creation story that implies only man was created in the image of God, and that woman was responsible for humanity's fall from grace.
2. Pauline passages (e.g., 1 Cor 11: 2–16, 14: 34–5; 1 Tim 2: 11–14, 5: 13–15; Col 3: 18–19; Eph 5: 21–3) which subordinate women to men.
3. St Augustine's contention that woman by herself is not the image of God (Ruether, 1983).
4. The disparaging views on sex and women held by early Church fathers like Tertullian and Jerome (Ranke-Heinemann, 1990).
5. Thomas Aquinas's thesis that women are 'defective and misbegotten' (Ruth, 1980, pp. 98–100).

6. The Church's historical teaching – based on a literal interpretation of the Bible and the doctrines of its revered Fathers – that women must not aspire to priesthood or holy offices (Bernard, 1989).
7. The Church's overall tendency to defend the *status quo* and to 'teach' rather than to 'dialogue' sincerely with its critics (McCormick, 1989, p. 72).

Not all the 'anti-women' quotations catalogued in feminist publications – at times out of context – do justice to the somewhat nuanced views of the Church on women. Few ecclesiastical leaders today would defend sexism by quoting from ancient writings that women find offensive. The ecclesiastical luminaries of the past were probably biased against women, but so were the intelligentsia of the past (see Clark, 1999). Currently, though, most liberal theologians, bishops and priests are sympathetic to the cause of feminists (see McCormick, 1989; Treston, 1989a, 1989b; Trautman, 1997), and priests are reported to be using maternal and androgynous symbolism (e.g., 'God's womb', 'God our Father and Mother') in liturgies, even without any ecclesiastical authorization (Finley, 1992, p. 186). Recent popes assert explicitly that the sexes are equal, as did Pope Pius XII in 1957: 'Man and woman are the images of God; in their own manner they are equal in dignity and have the same rights. It cannot be claimed in any way that woman is inferior' (quoted in Guillemette, 1989, p. 131). The new Canon Law of 1983 too 'rejects the antiquated notion of women's subordination on the basis of the creation accounts, and the idea that women are potential temptresses' (Bernard, 1989, p. 18).

Feminist theologians find the conciliatory gestures of the official Church far from satisfactory, for they still see the dominance of patriarchy, especially in the denial of priesthood and other hierarchical positions to women. Their goal of dismantling patriarchy, starting from the male gender of God, therefore, still remains unaltered.

Feminist alternatives

While almost all feminists are agreed that the male God propounded by patriarchy must be sacrificed at the altar of sexual equality, they do not agree on the alternatives. The alternatives range from giving up the entire Judeo-Christian religiosity to making minor changes in the language used to refer to God.

According to Goldenberg, 'Jesus Christ cannot symbolize the liberation of women. A culture that maintains a masculine image for the highest divinity cannot allow its women to experience themselves as the equals of its men' (1979, p. 22). She hopes to replace *theology* with *thealogy* (*thea* being 'goddess') and sees 'an image emerging that many women can live with', namely, 'the Goddess of feminist witchcraft' (p. 89).

B. G. Walker, a self-labelled sceptical feminist, feels convinced that the 'reason a feminist needs a skeptical view of father religion is the product of that religion and will remain so long as God is assigned a masculine gender' (1987, p. 19). True feminism, according to her, 'implies and embraces a return to the idea of the supreme Goddess, once nurturant Mother of everything including all gods, who owed her their allegiance and respect as the author of their being' (p. 19). Others who favor the worship of goddesses or of a supreme Goddess include Goldenberg (1979), Starhawk (1979), Fiorenza (1992b), and Christ (1992, 1997).

R. Gross favours an androgynous God: 'I propose to move beyond God the Father ... to an imagery of bisexual androgynous deity by reintroducing the image of God as female to complement the image of God as male. I wish to argue for this option because I am convinced that ... theism ... requires anthropomorphism' (1992, p. 168). Haddon favours worshipping God as 'God-Feminine' and 'God-Masculine' rather than as 'Goddess' and 'God' since the latter terms 'call to mind the image of two deities, rather than complementary facets of the One' (1988, p. xiv). She encourages the incorporation of Eastern non-Christian symbols and traditions, and is profuse in offering unorthodox images, metaphors and symbols for prayer and meditation.

Sister Elizabeth Johnson (1992), the author of *She Who Is*, is comparatively orthodox and argues in unwavering terms for the creation, development and propagation of female metaphors for God. She sees no objection to the use of male metaphors, but believes that it is imperative to use female metaphors – underdeveloped and secondary though they may be – in order to undo the damage patriarchy has done and to arrive at more theologically sound metaphors. Most moderate feminists and liberal theologians would favour the balanced views of Johnson. For example, among the ten theses put forward by Catholic women in German-speaking countries (Austria, Switzerland and Germany) in 1989 is the following: 'Language is an expression of consciousness and reality. The church's language oppresses women (brotherliness, sons of god ...). Thus a language is desired that names and makes them visible (sisterliness, sons and daughters ...)' (Lissner, 1991, p. 340).

Inclusive language

Although some feminists include in their agenda the total destruction or transformation of Judeo-Christian religions, not many churches seem pressed as yet to decorate places of worship with icons of a female-Christ, statues of provocative goddesses and monstrous divinities, or artifacts of witchcraft. Perhaps the feminist innovation that affects most believers at present is inclusive language, although even this, according to a survey, is thought to be peripheral or unnecessary by the majority of Catholics (see Hughes, 2002). Inclusive language, which even the secular press made light of as recently as the 1970s (see, e.g., Kanfer, 1972), has now become the norm in society and, naturally, is very familiar to most of the population, including Christian believers. Reformist and sensitive ecclesiastical authorities, biblical scholars, liturgical coordinators and university chaplains have already come up with biblical translations, prayers, hymns and liturgical books that are in inclusive language (Smith, 1994; Newman, 1995; Gold, 1995; *Guidelines*, 2001; Wisnesky, 2001). The International Commission on English in the Liturgy (ICEL), consisting mostly of academic scholars but supervised by Catholic Bishops, has revised several liturgical texts in inclusive language (Filteau, 1992).

There are, however, serious theological problems in adopting inclusive language, especially the so-called *vertical* inclusive language – concerned with gender references to God (see Trautman, 1997; Urda and Sabalaskey, 1999). It is not a simple matter of changing phrases like 'God the Father' to 'God the Father-Mother' since each change may affect the dogmatic content of faith and require reworking of all related texts and traditions that have existed for many years. Calling God by the non-sexist 'It', for example, would be anathema to Judeo-Christian religions since their God is a personal God *par excellence*, unlike the impersonal Brahman of Hindus, who may be referred to as It.

Theologians and biblical scholars agonize over vertical inclusive language, and the issue has become highly controversial and divisive (see Schneiders, 1986; Mankowski, 1994; Hitchcock, 1995; Marlowe, 2001). The Vatican has admonished Catholics to be wary of texts like the New Revised Standard Version of the Bible that are written in inclusive language (Gilchrist, 1995) and has called on ICEL to eliminate inclusive language from liturgical texts such as the Missal and the Lectionary of the Mass that ICEL had drafted in inclusive language (Hitchcock, 2000a, 2000b). In May 2001, the Congregation for Divine Worship and the Discipline of the Sacraments in Rome released a strictly worded document *Liturgiam Authenticam*, outlining the principles that should guide vernacular translations of Roman Catholic liturgical texts. The document, highly sensitive to contemporary values of equality, assures readers that a right interpretation of the texts 'excludes any prejudice or unjust discrimination on the basis of persons, gender, social condition, race or other criteria, which has no foundation at all in the texts of the Sacred Liturgy'. However, it is also extremely cautious about inclusive language, and debars any innovation regarding the gender of God: 'In referring to almighty God or the individual persons of the Most Holy Trinity, the truth of tradition as well as the established gender usage of each respective language are to be maintained.' While the traditionalists rejoice at such an intervention from Rome, the liberals find the document too suffocating and rigid (see Jensen, 2001; Moloney, 2001; Hitchcock, 2003, 2004).

The final versions of several Catholic liturgical texts are still being processed. Indications are that the new texts, when approved, will not be in inclusive language – to the discontent of feminists and liberals.

Conclusion

In their introduction to *Womanspirit Rising,* Christ and Plaskow observe astutely that feminists who challenged religious sexism 'found that nothing aroused the ire of male theologians and churchmen so much as the charge that traditional language about God is sexist. The question seemed to challenge the fundamental core of biblical revelation' (1992, p. 4). The reaction of theologians and churchmen is quite natural since altering the core of what is ordinarily taken to be biblical revelation and Christianity is in fact one of the declared goals of feminism. 'Language about God in female images not only challenges the literal mindedness that has clung to male images in inherited God-talk ... But insofar as "the symbol gives rise to thought," each speech calls into question prevailing structures of patriarchy' (Johnson, 1992, pp. 5–6). The gender of God is a volatile issue precisely because it shakes the foundations of what many Judeo-Christians, men and women, hold to dearly as true faith.

Only time can tell whether the Christian God-talk will undergo radical changes and cause corresponding changes in theology and ecclesiastical practices. Regardless of the validity and correctness of the feminist critique of religious language, the inchoate efforts of feminists to redefine Christianity might ultimately lead to the purification of religions and a better understanding of God. Meanwhile, less dogmatism and more openness on the part of both feminists and anti-feminists may help ease the tension that currently prevails between them.

Notes

1 The problems in God-talk discussed in this chapter may very well apply to religions other than the Judeo-Christian ones. The focus, however, is the God of

Judeo-Christian religions, especially of Roman Catholicism. At present the Judeo-Christian women are, perhaps, the most educated and most outspoken of all women, and their literary output is substantial, influential and widespread. The androcentrism, sexism and patriarchy of other religions have not yet been as thoroughly investigated, though they do exist (see, e.g., Jayawardena, 1986; Hekmat, 1997; Pandey, 2004). It would be erroneous to infer from this chapter that Judeo-Christian religions are in any way more sexist or more patriarchal than other religions.

2 Both *traditionalist* and *feminist* are terms with a variety of good and bad connotations (Carlin, 1992). In this chapter, *traditionalist* is used solely to denote one who defends traditions, especially the tradition of using male metaphors for God. Likewise, *feminist* is used to denote one who accepts the basic feminist concepts of sexism, androcentrism and patriarchy, regardless of other ideological differences. Neither traditionalist nor feminist is used here in any pejorative sense.

References

Bernard, F. (1989). Do women lack rights in the Church? *Theology Digest* 36(1), 15–8.

Britto, F. (1988). Effects of feminism on English. *Sophia Linguistica* 26, 139–49.

Buckley, G. A. (1991, May). Mary, the alternative to feminism. *Homiletic and Pastoral Review*, 11–16.

Carlin, Jr., D. R. (1992, 17 January). I'm not a feminist, but ... . *Commonweal*, 8.

Christ, P. C. (1992). Why women need the goddess? Phenomenological, psychological, and political reflections. In P. C. Christ and J. Plaskow (Eds) (pp. 273–87).

Christ, P. C. (1997). *Rebirth of the goddess: Finding meaning in feminist spirituality*. Reading: Addison-Wesley.

Christ, P. C., and Plaskow, J. (Eds). (1992) (1st edn 1979). *Womanspirit rising: A feminist reader in religion*. New York: Harper Collins.

Clark, B. (Ed.) (1999). *Misogyny in the western philosophical tradition: A reader*. New York: Routledge.

Collins, S. (1992). Theology in the politics of Appalachian women. In P. C. Christ and J. Plaskow (Eds) (pp. 149–58).

Daly, C. B. (1998, 14 January). Catholic church and women. *The Irish Times, Letters*. Retrieved 25 July 2004, from http://www.its.caltech.edu/~nmcenter/women-cp/urlychu2.html

Daly, M. (1973). *Beyond God the father: Toward a philosophy of women's liberation*. Boston, MA: Beacon Press.

Daly, M. (1992). After the death of God the father: Women's liberation and the transformation of Christian consciousness. In P. C. Christ and J. Plaskow (Eds) (pp. 53–62).

Emswiler, S. N., and Emswiler, T. N. (1974). *Women and worship: A guide to non-sexist hymns, prayers and liturgies*. New York: Harper and Row.

Ferreira, C. R. (1987, May). The feminist agenda within the church. *Homiletic and Pastoral Review*, 10–21.

Filteau, J. (1992, 29 May). Mass texts: Inclusive revisions proposed. *St. Louis Review* 51(22), 1 and 8.

Finley, M. (1992, 26 September). Male and catholic in confusing times. *America* 167, 186–7.

Fiorenza, E. S. (1983). *In memory of Her: A feminist theological reconstruction of Christian origins*. New York: Crossroad.

Fiorenza, E. S. (1992a). Woman in the early Christian movement. In P. C. Christ and J. Plaskow (Eds), (pp. 84–92).

Fiorenza, E. S. (1992b). Feminist spirituality, Christian identity, and Catholic vision. In P. C. Christ and J. Plaskow (Eds), (pp. 136–48).

Fiorenza, E. S., and Collins, M. (Eds). (1985). *Women: Invisible in church and theology* (= *Concilium 182*). Edinburgh: T. and T. Clark.

Frank, F., and Anshen, F. (1983). *Language and the sexes*. Albany: State University of New York Press.

Gilchrist, M. (1995). Vatican says 'no' to NRSV translation – 'Inclusive language' setback. Retrieved 25 July 2004, from http://www.ad2000.com.au/articles/ 1995/feb1995p14_791.html

Gold, V. R. (Ed.). (1995). *New Testament and Psalms: An inclusive version*. New York and Oxford: Oxford University Press.

Goldenberg, N. (1979). *Changing of the gods: Feminism and the end of traditional religions*. Boston, MA: Beacon Press.

Gornick, V., and Moran, B. K. (Eds). (1971). *Women in sexist society*. New York: New American Library.

Gross, R. (1992). Female God language in a Jewish context. In P. C. Christ and J. Plaskow (Eds) (pp. 167–73).

Guidelines for writing papers and essays. (2001). Santa Clara University: Graduate Program in Pastoral Ministries. Retrieved 22 July 2004, from http://www.scu.edu/academic/programs/pm/resources/paperguide.pdf

Guillemette, N. (1989). St. Paul and women. *East Asian Pastoral Review* 26(2), 121–33.

Guimarães, A. S. (1999). Is God mother? Background of a pontifical statement. *The Remnant*, 15 October 1999. Retrieved 23 July 2004 from http://www.traditioninaction.org/HotTopics/a009ht.htm

Haddon, G. P. (1988). *Body metaphors: Releasing God-feminine in us all*. New York: Crossroad.

Halkes, C. (1981). Themes of protest in feminist theology against God the father. In J. B. Metz and E. Schillebeeckx (Eds) (pp. 103–10).

Hekmat, A. (1997). *Women and the Koran: The status of women in Islam*. Amherst, NY: Prometheus Books.

Hitchcock, H. H. (1995). *Feminism and the language wars of religion*. Retrieved 21 July 2004 from http://www.adoremus.org/FeminismLanguage.html

Hitchcock, H. H. (2000a). Vatican's latest move to correct translation problems – ICEL needs 'thoroughgoing reform'. *Adoremus Bulletin Online Edition* 5(10). Retrieved 25 July 2004 from http://www.adoremus.org/2–00-ICEL.html

Hitchcock, H. H. (2000b). Vatican: ICEL Psalter 'a danger to faith'. *Adoremus Bulletin Online Edition* 6(2). Retrieved 25 July 2004 from http://www.adoremus.org/0400-ICEL.html

Hitchcock, H. H. (2003). New ICEL statutes signal coming changes: Unusual meeting prepares way for Missal in English. *Adoremus Bulletin Online Edition* 9(8). Retrieved 25 July 2004 from http://www.adoremus. org/1103ICEL.html

Hitchcock, H. H. (2004). Roman Missal translation update: Bishops receive ICEL Missal texts; translation norms. *Adoremus Online Edition* 10(1). Retrieved 21 July 2004 from http://www.adoremus.org/0304TranslationNorms.html

Hole, J., and Levine, E. (1980). Historical precedent: Nineteenth-century feminists. In S. Ruth (Ed.), *Issues in feminism: A first course in women's studies*. Boston, MA: Houghton Mifflin (pp. 463–71).

Hughes, A. C. (2002, 5 May). Words have consequences. *Archbishop Hughes Column – Clarion Herald*. Retrieved 26 August 2004 from http://www.archdiocese-no.org/words_have_consequences.htm

Hurcombe, L. (1987). *Sex and God: Some varieties of women's religious experience*. London: Routledge & Kegan Paul.

Inclusive language lectionary, An. (1983–5). Philadelphia: Westminster Press.

Jayawardena, K. (1986). *Feminism and nationalism in the third world*. London: Zed Books.

Jensen, J. (2001, 13 August). *Liturgiam authenticam* and the New Vulgate. *America*, 11–23. Retrieved 22 July 2004 from http://cba.cua.edu/novav.cfm

Johnson, E. A. (1984). The incomprehensibility of God and the image of God male and female. *Theological Studies* 45(3), 441–65.

Johnson, E. A. (1989). Mary and the female face of God. *Theological Studies* 50(3), 500–26.

Johnson, E. A. (1992). *She who is: The mystery of God in feminist theological discourse*. New York: Crossroad.

Kaczor, C. (1992, April). Inclusive language and revealed truth. *Homiletic and Pastoral Review*, 16–20.

Kanfer, S. (1972, 23 October). Sispeak: a Msguided attempt to change herstory. *Time*, 30.

Kaufman, I. (1992, April). A theology of woman. *Homiletic and Pastoral Review*, 63–5.

Kelly, C. (Ed.). (1992) *The enemy within: Radical feminism in the Christian Churches*. Milton Keynes: Family Publications.

Kramarae, C. (1981). *Women and men speaking*. Rowley, MA: Newbury.

Kreuzer, S. (1990). God as mother in Hosea 11? *Theology Digest* 37(3), 221–6.

LaCugna, C. M. (1992, 10 October). Catholic women as ministers and theologians. *America* 167, 238–48.

Lakoff, R. (1975). *Language and woman's place*. New York: Harper.

Lissner, A. (1991). Women in society and church: Ten theses. *Theology Digest* 38(4), 339–44.

Liturgiam Authenticam. Retrieved 21 July 2004 from http://www.vatican.va/roman_curia/congregations/ccdds/documents/rc_con_ccdds_doc_20010507_liturgiam-authenticam_en.html

Mankowski, P. (1994). Silk purses and sow's ears: 'Inclusive language' comes to mass. *Voices* Online Edition. Retrieved 26 August 2004 from http://www.wf-f.org/MankoSilk.html

Marlowe, M. D. (2001). *The gender-neutral language controversy*. Retrieved 20 July 2004 from http://www.bible-researcher.com/inclusive.htm

Martin, F. (1991). Feminist hermeneutics: An overview. *Communio: International Catholic Review* (Summer) 18(2), 144–63 [part 1], (Fall) 18(3), 398–424 [part 2].

Martin, J. H. (1987). The injustice of not ordaining women: A problem for medieval theologians. *Theological Studies* 48(2), 303–16.

Martyna, W. (1980). The psychology of the generic masculine. In S. McConnell-Ginet, R. Borker and N. Furman (Eds), *Women and language in literature and society*. New York: Praeger (pp. 67–78).

McCormick, R. A. (1989). *The critical calling: Reflections on moral dilemmas since Vatican II*. Washington, DC: Georgetown University Press.

McLaughlin, E. (1992). The Christian past: Does it hold a future for women? In P. C. Christ and J. Plaskow (Eds) (pp. 93–106).

Metz, J. B., and Schillebeeckx, E. (Eds). (1981). *God as Father? (= Concilium 143)*. New York: Seabury.

Millet, K. (1970). *Sexual politics*. New York: Ballantine Books.

Moloney, G. (2001). *On liturgy and language*. Retrieved 22 July 2004, from http://www.redemptoristpublications.com/reality/julaug01/editorial.html

Morton, N. (1992). The dilemma of celebration. In P. C. Christ and J. Plaskow (Eds) (pp. 159–66).

Murphy, P. F. (1992, 23 September). Let's start over: A bishop appraises the pastoral on women. *Commonweal*, 11–5.

Newman, B. M. (Ed.). (1995). *Holy Bible: Contemporary English version*. New York: American Bible Society.

O'Brien, E. (1964). *Varieties of mystic experience*. New York: Holt, Rinehart & Winston.

O'Faolain, J., and Martines, L. (Eds). (1973). *Not in God's image: Women in history from the Greeks to the Victorians*. New York: Harper Colophon Books.

Ostling, R. N. (1992, 23 November). The second reformation. *Time* 140(21), 42–8.

Pandey, R. (2004). Medieval experience, modern visions: Women in Buddhism. *Monumenta Nipponica* 59(2), 223–44.

Ranke-Heinemann, U. (1990). *Eunuchs for the kingdom of heaven*. Trans. Peter Heinegg. New York: Doubleday.

Responses to *Liturgiam Authenticam*. (2001). *Adoremus Bulletin* Online Edition 7(4). Retrieved 22 July 2004, from http://www.adoremus.org/0601responses.html

Ruether, R. R. (Ed.). (1974). *Religion and sexism: Images of women in the Jewish and Christian traditions*. New York: Simon & Schuster.

Ruether, R. R. (1983). *Sexism and God-talk: Toward a feminist theology*. Boston, MA: Beacon Press.

Ruether, R. R. (1985). *Womanguides: Readings toward a feminist theology*. Boston, MA: Beacon Press.

Ruth, S. (Ed.). (1980). *Issues in feminism: A first course in women's studies*. Boston, MA: Houghton Mifflin.

Santini, S. (2001). *The gender of the Holy Spirit: On the orthodox revision of the gender of the Holy Spirit*. Retrieved 26 July 2004 from http://www.geocities.com/kibotos2002/feminine.html

Schneiders, S. (1986). *Women and the Word: The gender of God in the New Testament and the spirituality of women*. New York: Paulist Press.

Smith, C. R. (Ed.). (1994). *The inclusive New Testament*. Brentwood, MD: Priests for Equality.

Smith, P. M. (1985). *Language, the sexes, and society*. New York: Blackwell.

Spender, D. (1985). *Man made language*. 2nd edn (1st 1980). London: Routledge & Kegan Paul.

Starhawk. (1979). *The spiral dance: A rebirth of the ancient religion of the great goddess*. San Francisco: Harper & Row.

Starhawk. (1992). Witchcraft and women's culture. In P. C. Christ and J. Plaskow (Eds), (pp. 259–68).

Steichen, D. (1985). From convent to coven: Catholic neo-pagans at the witches' sabbath. *Fidelity* 5(1), 27–37.

Steichen, D. (1992). *Ungodly rage: The hidden face of Catholic feminism*. Ft. Collins, CO: Ignatius Press.

Stone, M. (1992). When God was a woman. In P. C. Christ and J. Plaskow (Eds) (pp. 120–30).

Swindler, A. (1972). *Woman in a man's church: From role to person*. New York: Paulist Press.

Trautman, D. W. (1997). *Liturgical and biblical texts for the third millennium: The revised Sacramentary and Revised New American Bible Lectionary*. Retrieved 19 July 2004, from http://cba.cua.edu/millenn.cfm

Treston, K. (1989a). Creation spirituality and feminine consciousness. *East Asian Pastoral Review* 26(2), 150–5.

Treston, K. (1989b). Five loaves and two fishes: Feminist hermeneutics and biblical theology. *Theological Studies* 50(2), 279–95.

Urda, J. A., and Sabalaskey, B. (1999). *God the Father, the Trinity and vertical inclusive language*. Retrieved 19 July 2004, from http://www.ourladyswarriors.org/articles/inclusive.htm

Vasquez, A. (1989). Women in the history of the Church: Invisible and inaudible. *East Asian Pastoral Review* 26(2), 108–20.

Walker, B. G. (1987). *The skeptical feminist: Discovering the virgin, mother and crone*. San Francisco: Harper Collins.

Wisnesky, R. (2001). Religious leaders promote use of gender-neutral pronouns for God. *Stanford Review* 27(4). Retrieved 21 July 2004, from http://stanfordreview.org/Archive/Volume_XXVII/Issue_4/News/news1.shtml

3
Asymmetries of Male/Female Representation in Arabic

Samira Farwaneh

The correlation between language and gender has advanced to the forefront of sociolinguistic research, particularly after the widely acclaimed yet highly controversial work of Lakoff (1975) in which she examines societal gender inequity and its effect on linguistic performance, focusing on language used by women as well as language used to refer to women. She sheds light on several linguistic domains where gender asymmetry figures prominently. Her work generated an upsurge in research on language and gender issues, primary among which is the detection of sexist language usage in English and other languages. Evident throughout is the assertion that sexist language usage, that is, the overt expression of gender bias, manifests itself in a variety of linguistic domains: syntax, semantics, discourse and the lexicon, as demonstrated unequivocally in the study of naming practices, terms of address usage, sex language and use of metaphors (Pauwels, 1998). According to these studies, gender inequity manifests itself in a variety of patterns ranging from the subtle to the profound. These manifestations include the following:

1. The generic use of the masculine pronoun; for example, 'to each his own'.
2. Ordering the masculine form before the feminine; for example, 'husband and wife' rather than 'wife and husband'.
3. Euphemisms; such as using the term 'lady' instead of 'woman'.
4. Semantic derogation; for example, 'mistress' which is no longer the exact equivalent of 'master'.
5. Lexical and paradigmatic asymmetries; for example, the term 'spinster' which refers exclusively to females has no masculine counterpart in the English lexicon; nor does the title of address 'Miss' which reflects differentiation on the basis of marital status.

A few studies have emerged in the last two decades targeting gender differentiation in Arabic; for example, Abdel-Jawad (1986, 1989) on Jordanian, Haeri (1987, 1992) on Egyptian, Abu-Haidar (1989) and Bakir (1986) on Iraqi, among others.[1] The aforementioned works approach gender differentiation in language from a purely linguistic dimension; observing, documenting and analysing the effect of gender as a social variable on linguistic variables, and situating their findings within sociolinguistic and linguistic theories. Religion is a social variable, among other variables, whose effect is detected and quantified and contrasted with other variables such as age, education and occupation.

This work continues this avenue of research, yet it departs from previous works in that it does not view gender solely as a social category that explains male/female speech differences, but rather as a component of individual identity whose affirmation or negation can be detected via linguistic signals. Further, I do not focus, as is the case in the aforementioned works, on the linguistic behaviour of males and females, and how gender differentiation is encoded through language use; rather, I focus on the components of males' and females' respective identities, and the asymmetric, often inequitable, expression of these components in language. The questions to be asked are, then, how do men and women identify themselves and refer to each other, and to what extent is this identification symmetric across the two genders?

If imbalance in gender indexing is detected, then an investigation into the linguistic or sociocultural factors which may have precipitated such imbalance is in order. Arabic presents a curious target of research when it comes to gender. It is on the one hand a language, like any other, which functions as a vehicle for transmitting sociocultural norms and stereotypes, and presents, through constant usage, means for perpetuating those stereotypes. On the other hand, Arabic secured for itself in the last 2000 years a unique status as the primary symbol of religious identity in the eyes of many Muslims; a status matched only by Hebrew. As such, gender inequity in Arabic can be viewed as a direct reflection of gender inequity inherent in the religion it symbolizes. Given the religious significance of Arabic, and since this volume focuses on the languages of religion, then it is natural to add to the aforementioned two questions a third one: to what extent is gender indexing imbalance, that is, gender gaps and asymmetries, inscribed in religious texts, and to what extent is Islam, compared to its monotheistic predecessors, responsible for generating and fostering gender inequity?

To answer all these questions, I undertake an investigation of the linguistic representation of females and males drawn from within the nominal and adjectival systems in Arabic[2] to identify types of gender-related asymmetries in the use and reference of identity labels. Such asymmetries are examined through consideration and evaluation of the morphological structure and semantic features within three linguistic variables: personal names, symbolic of primary (or basic as termed in Eid 2002) identity; titles, symbolic of volitionally acquired identity; and terms of address and reference, symbolic of superposed (i.e., not acquired volitionally) identity. In the course of the discussion, I will address the role of the Islamic religion, both in its faith and cultural sense, in this asymmetric encoding of gender identity through language.

The data collected for this purpose come from different sources: dictionaries, novels, the media, informal discussions with native Arabic speakers and the author's own native repertoire. A short overview of the morphological system of Arabic is given, focusing on the role of grammatical gender therein; thereby establishing the null hypothesis that in a gendered language, parity is expected in the nominal and adjectival system from which identification labels are drawn. Deviation from this hypothesis warrants an exploration and explanation. In the last part of the chapter, I undertake an examination of personal names, titles and terms of reference respectively, highlighting the different types of asymmetry within each category. It will be shown that the data collected reveal three patterns of asymmetry: avoidance, shift and incongruence, discussed generically below, and more specifically in the following sections.

Avoidance

Avoidance phenomena were first mentioned as viable and potentially productive targets of research in Ferguson (1997) who called for more studies of cases where speakers use, consciously or unconsciously, certain lexemes or phones while avoiding others, especially in dialect contact situations, where dialect differences are levelled in favour of certain forms but not others. In her extensive study of Arabic, Persian and English obituaries, Eid (1994, 2002) elaborates on the notion of avoidance by revealing an interesting trend in the obituaries (especially in earlier ones), namely that personal names of Muslim women in obituaries during certain decades show a drastic decline, a phenomenon she terms 'name avoidance'. Her findings corroborate Romaine's statement that 'one of the subtle forms of discrimination against women is that they are simply not mentioned at all' (Romaine, 2002, p. 154). Eid's work on obituaries is the only work that correlates avoidance to gender identity. She observes that during certain decades in Egypt, Muslim female names seem to be more visible at times of social and political stability, and invisible during turbulent times, that is, when national and religious identity are in conflict with gender identity.

Avoidance in this paper refers to the absence of one gender category in a pair, rendering gender patterns incongruent, which I take as a sign of invisibilization of gender identity; examples from English are 'virgin' and 'whore' with their exclusively feminine reference. Similarly, in Portuguese (Pauwels, 1998) the word [marito] 'husband' has no feminine equivalent *[marita],[3] instead the word [moxer] 'woman' is used to refer to wifehood. Why, all things being linguistically equal, do we not find counter-gender equivalents of such terms?

Shift

Semantic shift, usually derogation, refers here to cases where both gender forms exist but exhibit inequitable semantic connotations, usually with semantic derogation of the feminine; for example, 'master' versus 'mistress', where the feminine form started off on equal footing then degenerated over time (Romaine, 2002, p. 159).

Incongruence

Incongruence refers to cases where the grammatical and referential features of the lexeme are not congruent, that is, the lexeme appears without the feminine marker /-a/ giving the appearance of a morphologically zero-marked masculine form, yet exclusively denotes a feminine referent.

As the chapter reveals these asymmetries, it will aim to relate them to their triggering socio-cultural influences, focusing on the role of religion, which is most often held responsible for gender inequity.

Morphological gender in Arabic

The salient characteristic of Arabic (and Semitic) morphology is that all lexical categories are derived from a bi-, tri- or quadriliteral root, which carries the core meaning of the word. Intervening vowels – vocalic melody – serve as functional morphemes indicating the number and class in the case of nouns and adjectives, or measures in the case of verbs and corresponding verbal nominals. Since identification labels are derived from nouns and adjectives, therefore every term of reference or identification, be it a personal name, a family name, a title or other, carries the semantic features of the root. Grammatical gender in Arabic, as in many Romance languages, is derived by a productive affixation rule as in the kinship terms in Table 3.1a. Masculine forms are derived by zero affixation, that is, they are identical to the bare stem. Feminine terms on the other hand are derived by appending the suffix /-a/ to the bare stem. Other kinship term pairs are arbitrarily marked for gender, that is, specified as [+feminine] or [−feminine], lexically,[4] as in Table 3.1b.

By default, gender is assigned on the basis of semantics; that is, nouns or adjectives denoting a male referent are assigned masculine gender, and vice versa. In such cases, there is an overlap between the morphological and referential features. If semantic gender cannot be discerned, as in the case of inanimate and deverbal nouns – analogous to gerundive nouns in English – then gender is assigned purely on a morphological basis; thus, forms ending in the morpheme /-a/ are assigned feminine gender, while zero-marked forms, that is, forms identical to the bare stem, receive masculine assignment. For example, deverbal nouns ending with /-a/, for example, [qiraa'-a] 'reading', [kitaab-a] 'writing' and ['idaar-a] 'management', require feminine agreement, whereas zero-marked deverbal nominals, for example, [xitaam] 'ending', [ibtisaam] 'smiling' and [taγriid] 'chirping of birds', take masculine agreement.

We will assume that the null hypothesis, stated in the introduction and to be tested in the following sections, stipulates that all masculine nouns and adjectives are appropriate identification and referential labels for males, and all feminine nouns and adjectives are appropriate for females. Yet an examination of personal, occupational and referential labels in Arabic reveals, even upon casual observation, a number of counter-examples. The following sections will take each variable in turn, revealing the imbalance in the representation of males and females, and examining the role of gender both as a social category, that is, 'the

Table 3.1 Kinship terms in Arabic

Masculine	Feminine	Gloss
(a) xaal	xaal-a	'uncle / aunt (maternal)'
'amm	'amm-a	'uncle / aunt (paternal)'
'ibn	'ibn-a	'son / daughter'
(b) rijjaal	mara	'man / woman'
walad	bint	'boy / girl'
'ax	'uxt	'brother / sister'
'ab	'um	'father / mother'

stereotypical assumptions about what are appropriate social roles for women and men' (Hellinger, 2002, p. 108), and a component of one's identity, that is, the way a person is perceived by oneself and others.

Asymmetries in personal names

Personal or first names are representative of the person's primary identity, hence the care taken by the parents or other name givers when choosing names for their offspring, thereby attempting to shape this identity. Although the name holder is not consulted in the personal name she or he is granted, personal names constitute the first cornerstone in the person's identity. Very few opt to volitionally change their first names, choosing instead to live fatalistically with the names they were given and its implications. The gender of the child, and the perception of gender values by the parents, is deterministic of the quality of names given to male and female children. Consequently, the domain of personal names is perhaps the best to exemplify the three aforementioned types of asymmetry: name avoidance, better described as name aversion, semantic shift which creates semantically inequitable gender terms, and incongruence, or the mismatch between grammatical and referential gender.

Since they serve as labels to identify human referents, names are derived from nominal and adjectival forms of different measures. Arabic with its grammatical gender categories provides the mechanism to derive both masculine and feminine forms from each noun or adjective. The expectation is, then, to find a feminine counterpart for every masculine personal name. This is indeed the case with many personal names, especially those derived from adjectival forms of an iambic CVCVVC template, or participial forms of the augmented trochaic template CVVCVC, as in the names in Table 3.2.

Table 3.2 Personal names in Arabic

Masculine	**Feminine**	**Gloss**
nabiil	nabiil-a	'noble'
samiir	samiir-a	'companion'
laTiif	laTiif-a	'gentle'
kariim	kariim-a	'generous'
saami	saami-a	'eminent'
baasim	baasim-a	'smiling'

Yet not all names can be so paired. We find gaps in the naming paradigm, not surprisingly, in the subdomain of religious names, although such names are also derived from adjectival (mostly passive participial) forms. As Abdel-Jawad (1986) points out, religious names derived from the roots /Hmd/ 'bless' and /'bd/ 'worship' are favored among Muslims, following the Prophet's saying that [xayru l'asmaa'i maa Hummida wa 'ubbid] 'the best names are the ones from the verbs Hamad and 'abad'. Although the Prophetic saying did not specify gender, such names are without exception (at least in the Arab World) bestowed upon male children only. Examples of common religious male-exclusive names from the aforementioned roots can be seen in Table 3.3.

The personal names in Table 3.3 hold religious significance, as they either signify the humans' relation to their creator, as in the /'bd/ derivatives, which are compound nouns with ['abd] compounded to one of the 99 attributes of God, or are variant appellations of Prophet Muhammad, as in /Hmd/ derivatives and the titles [muxtaar] and [muSTafa]. The restriction does not hold with respect to non-religious names from the same roots. One finds, for example, personal names like [Hamiid] and [Hamiida], [Hamdi] and [Hamdiyya] 'good'. This phenomenon exemplifies the first of our three asymmetries, namely avoidance or the absence of a morphologically possible gender-related term. What is the reason behind this avoidance?

Abdel-Jawad (1986) invokes taboo theory to explain this asymmetry, attributing it to a social taboo against endowing girls with the holiest of names. Such names or titles referring to Allah or the Prophet are considered too 'precious', in Abdel-Jawad's terms, to be bestowed upon females. One may argue that the morphological pattern of such names, that is, the passive participle form, renders them less desirable in the feminine, since an addition of the feminine marker augments the syllabic structure of the form thereby increasing its duration and difficulty of articulation.

Table 3.3 Religious names in Arabic

Masculine	**Feminine**	**Gloss**
'abdallah	*'abdatallah	'worshipper of God'
'abduh	*'abdatuh	'His worshipper'
muHammad	*muHammada	'blessed'
'aHmad	*Humdaa	'most blessed'
maHmuud	*maHmuuda[5]	'blessed'
muxtaar	*muxtaara	'chosen'
muSTafa	*muSTafaat	'favoured'

However, note the following names from the same morphological (passive participial) patterns: [maHbuuba] 'beloved', [mabruuka] 'blessed', which indicates that it is the religious significance of the name and not its morphological patterning that dictates its suitability for naming females. Thus, although Muslims generally hold the contention that all believers, be they males or females, are, or are supposed to be, Allah's worshippers and the Prophet's followers, only males can overtly bear the identity as such embodied in their first names. Avoidance of, or aversion to, feminine counterparts of religious names creates an inequitable gender representation in religious space, and consequently signals the exclusion of the female from this domain.

While avoidance is witnessed with respect to a number of religious names, only one religious name provides evidence of the second asymmetry, semantic shift, which ultimately and naturally leads to avoidance; that is the name [xaliil] 'beloved' and its missing feminine equivalent *[xaliila]. The masculine name is one of the titles of the prophet Abraham, referred to in the religious literature as [ibraahiim ulxaliil] 'Beloved of God'. Appending the feminine marker, however, yields the term [xaliila] which has been derogated to mean 'mistress' in the sense of an unlawful partner. This derogation renders the term ineligible as a naming label.

The third type of asymmetry, incongruence, is exemplified by compound names whose first member is a feminine noun, yet its referential feature, that is, its name bearer, is masculine. Compound nouns bear the gender and number feature of the head of the compound, the leftmost member of the compound. However, when feminine nouns like ['aay-a] 'evidence' and [Hujj-a] 'proof' are compounded to the form ['allaah] 'God', the compound is accorded a masculine identification which overrides the feminine feature of its first member. Hence, if the name acquires religious significance, it becomes male-exclusive. Such names have been promoted, especially among the Iranian Shiites, from a personal name to the rank of a title. The only feminine compound names which retain the feminine feature of the head of the compound are [hib-at allaah] and [ni'm-at allaah] 'gift of God'.

Exiting religious space into the realm of the secular, we find that secular names show the same gaps and semantic and featural inequities observed in religious names. Secular names, by virtue of their underlying roots, embody the attributes desired in a child by a name giver. These attributes correlate with the socio-cultural stereotypes associated with masculinity or femininity. The naming process in Arab culture is, therefore, a 'careful mental process that is influenced by linguistic, social, psychological, or cultural considerations' (Abdel-Jawad, 1986).

It is customary in many cultures to bestow upon girls personal names expressing beauty, peace, faith and elegance, stereotypically correlated with femaleness. Suffice it to ask yourself how many English-speaking boys or men, versus girls or women, you have met named Charity, Chastity, Faith or Hope. Arab cultures do not deviate from this trend. In Jordanian Arabic, for example, girls, not boys, are given first names that evoke notions of beauty, peace and piety (Abdel-Jawad, 1986). These gender-specific cultural norms are bound to precipitate gaps and asymmetries in the name paradigms, as we see below.

While the name [kaamil] 'perfect/complete' is quite common across the board in the Arab World, its feminine *[kaamila][6] is not. The name has an active participial or agentive nominal form, which is morphologically susceptible to feminization. Perhaps the aversion to conceptualizing perfection associated with femininity stems from the distorted interpretation of an aledged Prophetic Hadith (saying) that females are [naaqiSaatu 'aqlin wadiin] 'deficient in faith and mind' in the folk understanding of religion. A similar gap, which we interpret as avoidance, is found with the morphologically parallel name ['aadil] 'just' whose feminine counterpart *['aadil-a] is also unattested.

Incongruence, that is, the clash of morphological gender and the sex of the referent, is another consequence of cultural gender bias. When a common noun is transferred into a proper noun, the semantic features of the noun – regardless of its morphological gender – determines its eligibility as a male or female name. Thus, congruence between semantics and sex supercedes that of morphological gender and sex. The following female names in Table 3.4 are transferred from masculine common lexical and deverbal nouns.

Should congruence between morphology and reference be maintained, and the female names be given to males, the male name bearer

Table 3.4 Female names

Name	Gloss
'ibtisaam	'smiling'
yasmiin	'jasmine'
'ariij	'scent of flowers'
taγriid	'singing of birds'
wafaa'	'loyalty'
Hanaan	'tenderness'
'imaan	'faith'
xitaam	'end'

may be subjected to societal ridicule and accusations of effeminateness. The practice of name giving is one of the strategies that mulds, albeit unconsciously, the identity of the female to fit the dominant values of the culture. Palestinians, because of their tumultuous history and long experience with military occupation, diverge from this tendency and break the gender taboo by giving children of both sexes assertive masculine-marked first names evoking power and strength; for example, [kifaaH], [niDaal] and [jihaad] 'struggle'; incongruence still exists, but the underlying gender bias is eliminated.

Yassin (1986) states that name choice is prompted if not dictated by the operation of cultural, social, interpersonal and many factors distinguishable from the linguistic structure itself, which explains the lack of overlap between linguistic structure and the gender of names, but to what extent is religion part of these asymmetry-causing socio-cultural factors? Undoubtedly, Islam and Islamic tradition enriched the name repertoire and continues to influence the naming practice in the Arab World. Yet, it seems that religious tradition plays a dual function, both as a source and a deterrent; while it has been observed that male religious names endure in popularity and use, female religious names are either non-existent, for example, *[maHmouda], or are gradually falling out of use (Abdel-Jawad, 1986). On the other hand, secular names with religious themes, for example, ['afaaf] 'chastity, and ['imaan] 'faith', seem to be female-exclusive. As we move from personal names to titles, it will be shown that once women gain control over their own identification labels, the influence of religious tradition dissipates to leave ample room for gender egalitarianism.

Asymmetries in titles and terms of address

Titles and forms of address are indicators of the professional or social status of the addressee or referent. Unlike personal names, titles are acquired experientially and with volition. A title holder chooses to acquire a social or professional title through education, occupation or marriage. There is frequent overlap between titles and terms of address, since form of address usage reflects the addressor's recognition of the addressee's social and occupational status. In Arabic, titles and address forms precede the given (first) name, not last name, as is the case in English and other Western languages. The only difference between titles and forms of address is that the former is optionally definite. Forms of address, on the other hand, are obligatorily indefinite and are preceded

Table 3.5 Forms of address

(a) Title	id-doktoor-a nadya *or* the-doctor-F Nadia	daktoor-a nadya 'Dr Nadia'
(b) Term of Address	yaa doktoor-a nadya voc. doctor-F Nadia	*yaa id-daktoor-a nadya 'Oh Dr Nadia!'

Table 3.6 Titles

Masculine	**Feminine**	**Gloss**
(a) Professional Titles		
doktoor	doktoor-a	'Doctor / Professor'
bašmuhandis	bašmuhandis-a	'Engineer'
'ustaaz	'ustaaz-a	'Teacher'
(b) Social Titles		
'ax	'uxt	'Brother / Sister'
zamiil	zamiil-a	'Comrade'[7]
sayyid	sayyid-a	'Mr / Mrs'
*'aanis	'aanis-a	'Miss'

by the vocative particle /yaa/, as in example Table 3.5, which shows the difference between the two.

Professional titles are symmetric, with no paradigmatic gaps, which indicates that Arab women have gained equal access to many occupational fields that were previously male dominated. The examples in Table 3.6a demonstrate gender equity in occupational titles. Social titles, on the other hand, are relatively asymmetric, with female titles being binary, reflecting the marital status of a woman, while male titles are unitary, regardless of marital status, as in example Table 3.6b.

The lack of parity in social titles indicates the fundamental role of women's sexual status in constructing their gender identity in the eyes of society. This sexually-based asymmetry is not unique to the Arab/Islamic world, however. Other languages coined binary titles indicating women's sexual inexperience and hence eligibility for marriage, for example, English 'miss' versus 'Mrs', French 'Madame' versus 'mademoiselle', and Spanish/Italian 'señora' versus 'señorita/señorina' are a few examples.

To explain the parity of professional titles versus the incongruence of social titles, one has to consider closely the type of identity which these titles articulate. Titles represent a person's acquired identity (Eid, 2002); yet the way professional and social titles are acquired is not parallel. Professional titles are acquired volitionally, as a result of a person's wilful intent to acquire a professional status through upward education and employment. Social titles, on the other hand, are superposed on the person by others, as is the case with terms of reference discussed in the following section. Social titles are not endonyms, that is, terms of self-identification, never used by a person to refer to her/himself. You never hear a woman referring to herself as 'miss so-and-so', but many women are proud to announce their professional status by referring to themselves by their professional title. The equity of professional names, then, stems from the fact that they are identity labels chosen deliberately by women themselves to refer to themselves. Social titles are labels chosen by others to identify women, driven by the stereotypical misperception of gender and its correlation with age.

The limited number of address forms in contrast to referential terms leaves little room for asymmetry. While personal names discussed in the previous section and referential terms to be discussed in the following section exemplify all three types of asymmetry – avoidance, shift and incongruence – titles and address forms present only one instance of avoidance, namely the unpaired term ['aanisa] 'miss' and the absence or avoidance of its masculine counterpart. This is an interesting instance of avoidance, however. Unlike the religious names whose feminine counterparts are avoided out of taboo belief that religiosity and femaleness do not mix, avoidance in this case assumes a more positive quality. The term *['aanis] is avoided due to its negative implication for males. I will term this phenomenon 'positive avoidance'; a term is avoided if it carries negative values, hence its avoidance enhances or secures the addressee's positive status. As we shall see in the following section, avoidance of feminine terms, which I shall refer to as 'negative avoidance', aims to deny the addressee or referent an elevated status, and hence is parallel to semantic shift or derogation.

Asymmetries in referential terms

Referential terms, unlike terms of address, become a symbol of one's acquired identity without her/his knowledge, much less consent. They are superposed upon the referent by the speaker who is guided by his/her positive or negative perception of the referent. As such, these terms provide interesting evidence of societal gender-stereotypical perceptions. And, as is the case with personal names, provide abundant examples of avoidance, shift and incongruence.

Superficial parallelism

Let us first begin with paired referential terms, that is, terms exhibiting no gender gaps. Table 3.7 gives a few examples.

Table 3.7 Terms of reference

Masculine	Gloss	Feminine	Gloss
(a) zawj	'husband'	zawj-a	'wife'
(b) 'a'zab	'bachelor'	'azb-a	'bachelorette'
(c) 'ariis	'bridegroom'	'aruus	'bride'
(d) walad	'boy'	bint	'girl'
(e) muTallaq	'divorced'	muTallaq-a	'divorced'
(f) 'armal	'widower'	'armal-a	'widow'

Table 3.8 Examples

(a) nadya 'armala	'Nadia is a widow.'
nadya 'armalat kamaal Huseen	'Nadia is the widow of Kamal Hussein'
(b) kamaal 'armal	'Kamal is a widower.'
*kamaal 'armal naad ya Huseen	'Kamal is the widower of Nadia Hussein.'

These terms show perfect morphological and semantic symmetry. Each term has its gender counterpart, hence no avoidance, shift or incongruence. However, this superficial symmetry conceals disparaging usage if context is taken into consideration. This disparity applies particularly to the terms (d), (e) and (f). The term [bint] 'girl' and [walad] 'boy' represent different social dimensions: While the term [walad] is a purely age-related term referring to a male who has not reached manhood, the term [bint], on the other hand, is sexually related and refers to a woman, of any age, of no sexual experience, that is, a virgin. The terms for 'widow/widower' and 'divorced' in Table 3.7 (e)–(f) are referentially equivalent but contextually asymmetric: the feminine terms may be used in association with the woman's late or former spouse, thereby stating the spouse's name after the term as in Table 3.8a, while the masculine terms are used as pure referential terms stating the marital status of the man without association with his spouse as in Table 3.8b.

Asymmetries: positive and negative avoidance

We now turn to morphologically and semantically asymmetric terms that exhibit gender gaps, semantic non-equivalence or feature mismatch to illustrate the effect of avoidance, shift and incongruence on the use of referential terms. The terms in Table 3.9 show examples of unpaired gender terms, thus providing examples of avoidance. The starred forms indicate a gender gap in the paradigm.

The first two terms in Table 3.9 (a) and (b), are typical examples of gender inequity, whose equivalents are found in many languages. Virginity

Table 3.9 Term of reference asymmetries: avoidance

(a) *'a'dar	'adr-a	'virgin'
(b) *'aahir	'aahir-a	'whore'
(c) *'aqiil	'aqiil-a	'bondage, wife'
(d) ba'l	*ba'l-a	'master, husband'
(e) mu-Hallil	*mu-Hallil-a	'enabler'

and whoredom, which occupy the opposite poles of the sexual experience continuum, are culturally perceived as female-exclusive. Female virginity is a primary prerequisite for marriage, while male virginity is neither required nor desired. Sexual promiscuity, on the other hand, is observed, documented and penalized only if the agent is a female, hence the need to coin a feminine referential term for 'whore'; promiscuity among males goes unnoticed, and if noticed, escapes sanction; hence the absence of a masculine counterpart. Yet the tenets of Islam prescribe ['iffa] 'chastity and abstinence' to both sexes. Consequently, both masculine and feminine terms may be derived from the root /'f/ 'be chaste' as in ['afiif] and ['afiif-a] 'chaste'; both gender forms are used as personal names and referential terms. Interestingly, the term for 'whore' is not of native Arabic origin; its phonetic resemblance to its English cognate, and the juxtapositioning of two pharyngeals in the root /'hr/ which underlies the surface form, are indicative of its foreign origin. Native Arabic roots observe a co-occurrence restriction on adjacent phonemes of the same place of articulation (Greenberg, 1960).

The term ['aqiil-a] is used as term of reference identifying the woman as the wife of someone; therefore, it is always used in a construct phrase and never as a singleton noun. Some argue (Abdel-Jawad, 1989) that the term ['aqiil-a] may derive from the root /'ql/ meaning 'mind'; thus, to refer to a woman as the ['aqiila] of someone is to imply that she is the wise one in the marriage. Societal norms and practices in the Arab World, however, as well as the absence of the term outside the construct state, do not lend support to this interpretation.

Particularly intriguing are the last two terms in Table 3.9 (d)–(e) which are the only masculine terms with unattested feminine counterparts. The first term [ba'l] is borrowed from Phoenician and Hebrew and denotes ownership,[8] and is used in Arabic as an archaic form of address referring to a husband, particularly in lamentation (Badawi and Hinds, 1986, p. 88). The last form, [muHallil] 'enabler', is derived from the biconsonantal root /Hl/ 'to permit, to make kosher'. The surface form assumes the active participial form of measure II [mu-CaCCiC] reserved for causative function of lexical categories. This is the only term in the list deeply rooted in Islamic tradition, and whose maintenance is justified on purely religious grounds. According to all interpretations of Islamic law, divorce is revocable after the first or second pronouncement and may be abrogated within the three-month waiting period following each pronouncement. Divorce becomes absolute once the third repudiation is pronounced, whereupon the husband may not have conjugal access to his wife through a fourth marriage. However, in some interpretations, a husband may reclaim his irrevocably divorced wife if the wife marries a second husband, and is divorced after consummating the second marriage. She may then return to her first husband. The second husband is referred to as a [muHallil] or 'enabler'; his role, be it intentional or accidental, is to enable the first husband to reclaim his wife after a third repudiation. A search of the two primary sources of Islamic law, the Qur'an, Muslims' Holy Book, and Hadith, the sayings of Prophet Muhammad, yielded no instantiations of the term [muHallil], which indicates its latter-day coinage. The concept, however, is documented in a few Hadiths – prophetic sayings – in Sahih Bukhari, one of the collections of Hadiths. One such Hadith goes as follows:

> A man divorced his wife thrice (by expressing his decision to divorce her thrice), then she married another man who also divorced her. The Prophet was asked if she could legally marry the first husband (or not). The Prophet replied, 'No, she cannot marry the first husband unless the second husband consummates his marriage with her, just as the first husband had done.'
>
> (Volume 7, Book 63, Number 187)

Interestingly, the other main source of Hadith, Sahih Muslim, bears no mention of this concept, which sheds doubt on its authenticity. Authenticity not withstanding, our analytical interest lies in the absence of a feminine counterpart of the term, that is, *[muHallila]; like the term [ba'l], [muHallil] is another status term embodying authority and power grounded on religious superiority, and as such, is male-exclusive.

In sum, forms denoting subservience, sexual promiscuity or inexperience seem to be exclusively female with no masculine counterpart. This is a manifestation of 'positive avoidance' where a negatively connotated or powerless term is avoided to enhance or preserve the superior status accorded to the referent, here the male. Terms denoting power and authority, on the other hand, tend to be male-exclusive. Avoidance of a feminine equivalent is seen here as a form of 'negative avoidance', that is, avoidance of prestigious elevated forms to deny or demote the identity of the referent, here, the female.

Asymmetries: semantic shifts

The identification labels discussed in this subsection differ from the previously mentioned ones in that both gender forms are attested, but with disparaging connotations. The masculine form connotes prestige and status while the feminine form is belittling at best, if not outright derogatory. Consider the paradigm in Table 3.10. Down arrows indicate negative semantic shift.

What all the forms in Table 3.10 have in common is that appending the feminine morpheme /-a/ to the masculine form involves downward semantic shift ranging from marginalization to derogation. The form [rabb il-beet], for example, is demoted to a mere 'housewife' when the feminine suffix is appended. One may be tempted to argue that a [rabbat beet] is equivalent to a mistress of the house, an equally powerful term. However, as Lakoff (1975) and Romaine (2002) point out, the connotations of words do not arise from the form in isolation but from the context of their usage. Collocations 'transmit cultural meanings and stereotypes which have built up over time' (Romaine, 2002, p. 160). The feminine form usually collocates with negative constructions and minimizing modifiers such as 'just' or 'only' as in 'she knows nothing about the outside world because she's just a housewife', while the masculine term associates with positive declarative constructions as in 'he makes all the decisions because he is the master of the house'.

The referential terms in Table 3.10 (b)–(e) all indicate how terms may begin on an 'equal footing,' then the feminine term degenerates over time (Romaine, 2002, p. 160). According to Said, (1978), the phonetically non-reduced form ['aalima] used to refer to a learned woman who is an

Table 3.10 Semantic shift

Masculine	Gloss	Feminine	Gloss
(a) rabb il-beet	'master of house'	↓rabbat beet	'housewife'
(b) 'aalim	'scholar'	↓'alma	'entertainer'
(c) waliyy	'holy man'	↓wiliyya	'woman-pejorative'
(d) γaazi	'warrior'	↓γaziyya	'entertainer'
(e) šeex	'Sheikh'	↓šeexa	'entertainer-in North Africa'
(f) maHram	'travelling companion'	↓Hurma	'woman'

accomplished recitor of poetry and the Qu'ran. By the mid-nineteenth century, the term degenerated and the meaning shifted to become a reference to a female dancer and singer who may practice prostitution on the side (Said, 1978, p. 186). Similarly, the feminine forms in Table 3.10 (c)–(e) lose their original meaning and gradually acquire an abusive and pejorative connotation, often with implications of sexual promiscuity. Notice here that semantic shift toward pejoration is associated with reductive phonological processes: ['aalima] becomes ['alma] via vowel shortening and syncope (deletion of a word-internal vowel), [waliyya] is rendered [wiliyya] by vowel raising (which involves reduction in sonority), and [γaaziya] converts to [γaziyya] also by vowel shortening. The correlation between phonological reduction and pejoration becomes transparent once we consider the nature of pejoration. Pejoration has many social and psychological functions, one of which is to demote the status of an individual out of spite or in jest; another is to stereotype an individual for her membership in a social or ethnic group. Stereotyping, as defined in Talbot, involves 'simplification, reduction, and naturalization' (2003, p. 169). Phonological reduction, then, can be viewed as an auditory manifestation of social reduction.

Another dimension of stereotyping is objectification – association of a human referent to an inanimate object – and zoologization – association to an animate non-human object (Maalej, 2003); the last pair in the list [maHram] and [Hurma] is an illustrative example. Both are derived from the root /Hrm/ 'ban/prohibit' which is the antonym of /Hl/ 'permit' from which the term [muHallil] discussed in the previous section is derived. The masculine form [maHram] refers to a male travelling companion required of women otherwise travelling alone. This companion must be a male kin whose blood relation to the woman pre-empts marriage. The surface form assumes the templatic pattern ma-CCaC which identifies instrumental nouns, as in [mašraT] 'scalpel' and [maqla] 'fryer'. The form can, then, be interpreted literally as 'instrument of prohibition'. Contrast the authoritative connotation of the masculine form to its feminine counterpart [Hurma], used as a euphemism or derogation for 'woman'. The form takes the nominal templatic pattern CuCC-a reserved for inanimate nouns such as [lu'ba] 'toy', [jumla] 'sentence' and [rukba] 'knee'. [Hurma] is the only nominal of this pattern with a human referent. One may therefore interpret the form literally as 'object of prohibition'; the current use of the plural [Hariim] in both Eastern and Western writings attests to the inferior status the term allocates to women.

This section shows how gender inequity is reflected through semantic shift or derogation of the feminine. This shift yields an abundance of masculine terms denoting religiosity and chivalry versus feminine forms denoting promiscuity and worldliness with abusive connotations. This, however, is not unique to Arabic or Islam. Semantic derogation of these forms parallels that of English terms like 'lady', 'mistress' and 'courtesan', which indicates the universality of the tendency to belittle or marginalize the feminine through linguistic means.

Asymmetries: incongruence

Personal names provided examples of incongruence where the referential and morphological gender features do not match. These are usually of the deverbal noun form. Referential terms provide a yet more interesting example of incongruence. Few female referential terms appear in the masculine zero-marked form, that is, they do not inflect morphologically for gender, despite their referential feminine feature. A few examples are listed in Table 3.11.

The first term [Haram] is another derivative of the root /Hrm/ 'ban'. When used as a singleton noun, qualified or unqualified, it refers to a sacred enclave (Badawi and Hinds, 1986, p. 201); as in [ilHaram iššariif] 'the Grand Mosque of Makkah or Jerusalem', or [ilHaram iljaami'i] 'the sanctuary of the university'. As a first member of a construct phrase, for example, [Haram il'ustaaz] 'wife of the professor', the form is interpreted obligatorily as a female-exclusive label, despite the absence of feminine marking, signalling marital status. The converse [Taaliq] is another zero-marked form denoting a woman's marital status, or the reversal thereof.

The third term ['aanis], like its English counterpart, collocates with negatively connotated terms indicating plainness, old age and desperation (for marriage) (Romaine, 2002), and in the Arabic popular culture correlates with the profession of school teaching.

Another religiously-based gender term is represented in Table 3.11 (d). This referential term denotes a wife who oversteps the bounds of her marital role as prescribed by the husband. She is then deemed 'disobedient' and can hence be forced into the marital house, termed

Table 3.11 Incongruence

Term	Gloss
(a) Haram	'wife of'
(b) Taaliq	'divorced'
(c) 'aanis	'spinster'
(d) naašiz	'disobedient'

[bayt iTTaa'a] 'house of obedience'. This rule which made its way into the personal status law in many Arab countries derives its legitimacy from the following verse: 'Should you fear their (Fem) disobedience [nušuuz], admonish them, desert their beds, then beat them ...' (Qur'an Sura 4 (Surat an-nisaa' 'women'), aya 134).

Searching for other instances of the word [nušuuz], one finds another Aya (verse) in the same Sura (chapter), where the referent this time is the male:

> If a wife fears cruelty [nušuuz] or desertion on her husband's part, there is no blame on them if they arrange an amicable settlement between themselves; and such settlement is best.
>
> (Quran sura 4 (Surat an-nisaa' 'women'), aya 128)

Though the Qur'anic verses utilize the deverbal noun [nušuuz] in reference to a character deficiency observed in both genders, the civic law (enacted exclusively by men) coins a masculine participial form [naašiz] referring only to women.

The negative social implications of the forms are evident, and so is their morphological incongruence. The use of the masculine form when the referent is female cannot be attributed to the typical, much frowned upon, generic use of the masculine to refer to both sexes. Instead, the masculine form is used as the default when no gender opposition is assumed. Therefore, the term for 'pregnant' appears in most Arabic dialects in its masculine zero-marked form [Haamil] not *[Haamila]. One can therefore glean from the use of the masculine in these cases the societal misperception of the female as the only sex that can be owned through marriage [Haram], divorced without volition [Taaliq], spinsterly ['aanis], and disobedient [naašiz].

Conclusion

In this chapter I have outlined the dimensions of gender inequity by categorizing the types of asymmetry observed in the usage and choice of three linguistic variables: personal names, titles and address forms, and terms of reference. It has been shown that the masculine forms of stereotypical feminine terms are avoided to preserve the prestige of the masculine, while the feminine counterparts of elevated masculine terms are either avoided or derogated to deny or trivialize female identity. It was also shown that mismatches between semantic, morphological and referential features are attributable primarily to a stereotypical view of females and their role in society.

An attempt to trace the observed inequitable terms directly to the Qur'an has proven unsuccessful, and those purported to come from the Hadith either do not transmit from reliable sources, or lack consensus authentification. Hence, one may safely assume that the gender representational imbalance observed in Arabic is not due to Islam *per se*, but rather to the folk understanding of Islam, which views religious identity as antithetical to female visibility.

Once women claim ownership of their identity, the impact of folk religious tradition dissipates allowing women to acquire equitable representational visibility, as is evident from the parity of occupational titles. Should women claim similar ownership of their religious identity, through active participation in scriptural interpretation, the gap between gender and religious identity, which precipitates disparaging linguistic expressions, is bound to disappear.

Notes

1 Reasons for the deficit of language and gender research, especially within the geographic bounds of the Middle East, include the salience of national identity which often overshadows gender concerns, pseudo-religious traditions, state-prescribed limitations on academic freedom and the impact of traditional education on critical thinking and research.

2 The linguistic target here is Modern Arabic dialects. For the purposes of this chapter, however, I am not focusing on any particular variety. Should there be any noteworthy dialectal variation, it will be mentioned.

3 Similarly in French, two masculine terms exist, 'homme' and 'mari', referring to manhood and husbandhood respectively; womanhood and wifehood, on the other hand, converge under one surface form 'femme'.

4 That is, both forms are present in the lexicon, rather than deriving one from the other by suppletion. External evidence corroborating this claim comes from the absence of typical overgeneralization errors which characterize L1 and L2 speech, such as the production of *goed* for *went*. There is no evidence to date of overgeneralization such as the following, whereby a feminine marker is appended to the bare form: [walada] for 'girl', [rijjaala] for 'woman'. We do observe overgeneralizations with true suppletive plurals, where the L1 or L2 speaker may provide [*mart-aat] and [*HSan-aat] as the plural of [mara] 'woman' and [HSaan] 'horse', instead of the correct suppletive forms [niswaan] and [xeel].

5 An internet search of the name [maHmuuda] 'Mahmouda' yielded around ten results, all of which are ethnonymically and linguonymically non-Arab; for example, Afghanistan, Pakistan, Mauritania, Senegal and Kenya.

6 This name is common in both Ottoman and Modern Turkish (S. Kuru and N. Babur, personal correspondance).

7 Usually used in socialist countries like Syria and Iraq. It would be interesting to witness the fate of such socialist terms in Iraq, as the political map is being redrawn.

8 Ussishkin (personal correspondance) states that in Hebrew [ba'al] has a dual meaning depending on the term it is juxtaposed to. It denotes ownership if juxtaposed to an inanimate noun, for example, [ba'al bayt] 'owner of a house' or marital status when appended to a human noun or pronoun, for example, [ba'ala] 'her husband'. The feminine equivalent [ba'ala] is used only in the sense of ownership and never marital status; thus, [ba'alat bayt] but not *[ba'alato] 'his wife'.

References

Abdel-Jawad, H. R. (1986). A linguistic and sociocultural study of personal names in Jordan. *Anthropological Linguistics* 28(1), 80–94.

Abdel-Jawad, H. R. (1989). Language and women's place with reference to Arabic. *Language Sciences* 11(3), 305–24.

Abu-Haider, F. (1989). Are Iraqi women more prestige conscious than men? Sex differentiation in Baghdadi Arabic. *Language in Society* 18(4), 471–81.

Badawi, E. S., and Hinds, M. (1986). *A dictionary of Egyptian Arabic: Arabic–English*. Beirut: Librairie Du Liban; issued under the sponsorship of the American University in Cairo. Beirut, Lebanon.

Bakir, M. (1986). Sex differences in the approximation to standard Arabic: A case study. *Anthropological Linguistics* 28(1), 3–9.

Eid, M. (1994). 'What's in a name?': Women in Egyptian obituaries. In: Y. Suleiman (Ed.), *Arabic sociolinguistics: Issues and perspectives*. Richmond: Curzon Press (pp. 81–100).

Eid, M. (2002). *The world of obituaries: Gender across cultures and over time*. Detroit, MI: Wayne State University Press.

Ferguson, C. A. (1997). Standardization in Arabic. In: R. K. Belnap and N. Haeri, (Eds), *Structuralist studies in Arabic linguistics: Charles A. Ferguson's Papers, 1954–94* (Studies in Semitic Languages and Linguistics). Leiden: Brill.

Greenberg, J. (1960). The patterning of root morphemes in Semitic. *Word* 6, 162–81.

Haeri, N. (1987). Male/ Female differences in speech: An alternative interpretation. In: K. M. Deming et al. (Eds), *Variation in language: NWAV-XV at Stanford (Proceedings of the Fifteenth Annual Conference on New Ways of Analyzing Variation)*. Stanford, CA: Department of Linguistics, Stanford University (pp. 173–82).

Haeri, N. (1992). How different are men and women: Palatalization in Cairo. In: E. Broselow, M. Eid, and J. McCarthy (Eds), *Perspectives on Arabic linguistics IV*. Amsterdam and Philadelphia: John Benjamins (pp. 165–80).

Hellinger, M. (2002). Gender in a global language. In: M. Hellinger and H. Bußman (Eds), *Gender across languages: The representation of women and men*, vol. 1. Amsterdam and Philadelphia: John Benjamins (pp. 105–13).

Lakoff, R. (1975). *Language and woman's place*. New York: Harper and Row.

Maalej, Z. (2003). Metaphor and first names. ms, University of Manouba, Tunes.

Pauwels, A. (1998). *Women changing language*. London and New York: Longman.

Romaine, S. (2002). A corpus-based view of gender in British and American English. In: M. Hellinger, and H. Bußman (Eds), *Gender across languages: The representation of women and men*, vol. 2. Amsterdam: John Benjamins (pp. 153–75).
Said, E. (1978). *Orientalism*. New York: Pantheon Books.
Talbot, M. (2003). Gender stereotypes: Reproduction and challenge. In: J. Holmes and M. Meyerhoff (Eds), *Handbook of language and gender*. Oxford: Blackwell (pp. 468–84).
Yassin, M. A. F. (1986). The Arabian way with names: A sociolinguistic approach. *Linguistics* 25(2), 77–82.

4
American Women: Their Cursing Habits and Religiosity

Timothy Jay

Cursing is the use of offensive emotional language to express one's emotions and communicate them to others. Cursing is ubiquitous in American social life. Many questions remain to be answered since scholars have given the phenomenon scant attention over the years. Here I focus on two questions: Why do American men curse more than American women do?; What role does religion play in the process? These questions allow us to review research on three psychosocial factors, gender identity, religiosity and cursing. My aim is to demonstrate that people who curse use offensive language primarily to express anger or frustration, and that gender and religiosity moderate this habit (as do mental status, hostility and alcohol use). Religious women seem to be doubly restricted from cursing, first for their gender (men can express aggression more openly than women can) and second for their religious beliefs (Christians should not use profanity). I review religious restrictions on language, the relationship between religiosity and cursing, the relationship between sex anxiety and cursing, women's use of taboo language, and working-class women as a counter-example. I begin the discussion with an examination of recent increases in women's public cursing and an outline of cross-cultural cursing comparisons.

American women's cursing: past and present

Curse words persist over hundreds of years because they are useful to a culture (see Hughes, 1991). Until recently it was difficult to get an accurate measure of how frequently curse words were used in public. All written records and documents have been censored and estimates of cursing based on written materials are entirely unreliable. Frequency *estimation* techniques and recording methods are more accurate than written materials. Besides, cursing is primarily an oral, not a written phenomenon.

To provide data on the frequency and stability of cursing, Jay compared a field study of cursing from a sample from 1986 (see Jay, 1992) with a set of 1996 data recorded on the east and west coasts of the United States (Jay, 2000). The 1996 study permits geographic comparisons and time-frame comparisons. Data were collected by male and female researchers who recorded episodes of cursing in and around college communities in Los Angeles, California, Boston, Massachusetts and in western Massachusetts. Half of the data were recorded by females and half by the author. Half of the data were recorded in California and half were recorded in Massachusetts.

Most of the data from California (90 per cent), are accounted for by the usage of ten words. Two words *fuck* and *shit* account for 50 per cent of the cursing episodes. Most of these curse words are obscenities (*fuck, shit*) or profanities (*god, hell, damn*). Females were much more likely to say the mild oath 'Oh my god' than males. Males were recorded swearing more than females, 56 per cent and 44 per cent respectively. Both males and females generally use the same set of words with a few exceptions. Males had a production vocabulary of 28 words and females, 20. The correlation between male and female vocabularies is high, r = 0.75 to 0.80.

In Massachusetts most of the data, 90 per cent, are accounted for by the usage of ten words. *Fuck* and *shit* accounted for 54 per cent of the data. Most of the curse words are obscenities and profanities. Again, females were much more likely to say 'Oh my god' than males. In this sample males and females were recorded cursing at about the same rate. Males had a production vocabulary of 22 words and females, 24. The correlation between males and female cursing is quite high, r = 0.93.

The correlation between the most frequent words used on the east coast with their west coast counterparts is quite high, r = 0.97. This means that there is very little difference in cursing in these two different locations. One important difference is that females were recorded swearing more in public in the east, relative to the west coast. This might be due to the fact that only one female recorded data in the west, while three females recorded data in Massachusetts.

The past decade – 1996 and 1986

Most cursing involves the use of a small set of curse words that are repeated often. Not much has changed for public cursing during the last ten years. Speakers in a college community rely heavily on obscenity (*fuck, shit*) and profanity (*hell, Jesus, goddamn, damn, god*). Males tend to use more obscenities than females, who use more profanities than males. Interestingly, one finds the opposite emphasis (more profanity and few obscenities) in a nursing-home setting, where speakers in their eighties and nineties are less likely to utter strong obscenities (Jay, 1996). There are also more women in nursing homes than men. Overall, one might notice that extremely offensive language occurs at a low rate in public, words such as *cocksucker, cunt, nigger* or *spic* were heard infrequently around campus communities. One noticeable difference is that American women are swearing more in public than they did 20 years ago. We conclude that American cursing is fairly stable, involving a small set of words repeated frequently, mainly obscenities and profanities. The stability of these cursing patterns over ten years suggests that cursing in public has not undergone dramatic changes. Before moving on we need to address another question at this point. First, how does American cursing compare to other cultures?

Cross-cultural comparisons

One method for making cross-cultural comparisons is to look at patients with similar cursing problems across cultures (see Jay, 2000). For example, Tourette Syndrome (TS), a tic disorder associated with compulsive cursing (coprolalia), occurs in all cultures and there is uniformity in its clinical picture. What is missing in the picture is that a Touretter from a non-English-speaking country utters forbidden words in his or her culture, not what is forbidden in English. Because the coprolalic lexicons differ depending on culture, cross-cultural comparisons of TS lexicons reveal the semantics of forbidden words in a culture. Coprophenomena in TS indicate a neurological failure to inhibit thoughts and speech learned in childhood that are forbidden within the child's culture. Cross-cultural coprolalia reveals the universal use of religious, sexual, scatological and animal references. However, the relative frequencies of these references (religious versus sexual, for example) depend on culture. Whether a Touretter utters profanity or not depends on his or her culture.

Meaningful background information about TS appears in the work of Shapiro, A., Shapiro, E., Young and Feinberg (1988). As for cross-cultural comparisons of Touretters' lexicons, one of the first was made by Lees (1985). Lees made comparisons of Touretters' frequent coprolalia based on US, UK, Hong Kong and Japanese studies (Table 4.1). Several reports have surfaced since Lees's work, which are examined later.

Table 4.1 Comparisons of Touretters' coprolalia

United States	United Kingdom	Hong Kong	Japan
fuck	fuck	tiu (fuck)	Kusobaba (shit grandma)
shit	shit	shui (useless person)	chikusho (son of a bitch)
cunt	cunt	tiu ma (motherfucker)	(female genitalia and breasts)
mother-fucker	bastard	tiu so (aunt fucker)	

Note: Table adapted from Lees (1985); US data from Shapiro et al. (1978); UK data from Lees et al. (1984); Hong Kong data from Lieh Mak et al. (1982); Japan data from Nomura and Segawa (1982).

American, British and Canadian English

As Lees's work and that of others (Shapiro, A., Shapiro, E., Young and Feinberg, 1988) indicated, English coprolalia most frequently employs obscenities (*fuck, cocksucker, shit, cunt, motherfucker*) and socially offensive words such as *bitch, bastard* and *nigger*. Obscenities and socially offensive words predominate over milder profanities (*hell, damn, Jesus*). One theory is that obscenities relieve the stress associated with coprolalia more effectively than mild profanities.

Beyond the English-speaking world, one has to ask if non-English-speaking cultures produce coprolalia similar to English-speaking counterparts with TS? The answer is not straightforward. Some of the semantics underlying coprolalia, for example references to genitalia, religion, animals or faeces, are remarkably similar.

Middle East

One of the studies that makes it obvious how sensitive coprophenomena are to culture is Robertson and Trimble's (1991) analysis of five patients from the Middle East with TS. The most interesting case is a young woman born in Kuwait of an Arabic background. Her coprolalia began at the age of 15. The literal translations of the Arabic words were *ass, bitch* and *pimp*. But more telling was her sexual disinhibition in public, which included uncovering her thighs in public, unacceptable in Moslem culture, and exposing her breasts at school.

Japan

Several authors working on TS have stated that the disorder occurs only rarely in Asian cultures, referring mainly to Japanese and Chinese research. Nomura and Segawa (1982) reported a study of 100 Japanese

TS cases. According to the report, coprolalia is also infrequently seen in Japanese aphasics, but not uncommon in English-speaking aphasics. One caution, reported differences in prevalence may be due to comparing samples with different age ranges and not due to cultural/genetic differences.

Common Japanese words reported in Table 4.1 include *kusobaba* (*shit grandma*), an insult usually directed at an older woman, *chikusho* (*son of a bitch*), an animal reference to domesticated animals, comparable to expletives of frustration in English (*damn it*). Other words were references to female genitalia and breasts.

Brazil

Cardoso, Veado and de Oliveira (1996) studied the clinical features of 32 Brazilian patients (24 men and eight women) with TS. Coprolalia and copropraxia were present in nine patients. The lexicon of 'obscenities' shouted by the patients is as follows:

merda	*faeces*
bosta	*faeces*
filho da puta	*son of a whore*
bunda	*buttocks*
buceta	*vagina*
cacete	*penis*
caralho	*penis*
porra	*sperm*
va tomar no cu	*fuck off*

(Cardoso, Veado and de Oliveira, 1996, p. 210)

The authors suggest that coprolalia represents an expression of disinhibition and patients with TS become incapable of suppressing the production and vocalization of obscenities 'which vary depending on culture'.

Spain

Lees and Tolosa (1988) in a report on tics, listed the following curse words from Spanish patients with TS. The words are listed in order of frequency:

puta	*whore*
mierda	*faeces*

cono	*vulva*
joder	*fornicate*
maricon	*homosexual*
cojones	*testicles*
hijo de puta	*son of a whore*
hostia	*holy bread, literally*

Denmark

Regeur, Pakkenberg, Fog and Pakkenberg (1986) studied 65 patients with TS in Denmark, who were being treated with Pimozide for their symptoms. Seventeen patients exhibited coprolalia. Examples of their 'obscenities' (with the authors' translations, p. 792) included:

kaeft	vulgar expression for shut up
svin	swine – rather powerful in Danish
fisse, kusse	very vulgar expressions for the vulva
pik	vulgar expression for the penis
rov	ass
pis	piss
sgu	by God
gylle	rustic word for farm animal excretions
lort	shit

These examples seem somewhat similar to English coprolalia, in that they refer to body parts, genitalia and body products. The animal terms and profanity are less typical of English-speaking Touretters but still typical of English cursing.

Hong Kong

Lieh Mak, Chung, Lee and Chen's (1982) study of coprolalia is based on 15 Chinese patients, born in Hong Kong and treated there. The original report indicated that seven patients used single swear words (not reported) and two used phrases like 'fuck your mother' and 'rape your aunt'. Their families considered coprolalia to be the most undesirable symptom, but the patients did not seem to be distressed by the symptom. According to Table 4.1, Lees reported the patients making references to female genitalia, breasts and useless persons, but these words are not in the original report. And one has to wonder about the difference between 'aunt fucker' and 'rape your aunt' as translations. They seem to be meant as equivalent interpretations of a Chinese term, but are they? A caution on translation is necessary.

It is tempting to conclude that English speakers are more obscene and Brazilian, Danish and Spanish speakers are more religious because they present different patterns of coprolalia. Japanese speakers also produce less obscene and religious references, employing more ancestral allusions and insulting references. We can infer that profanities are salient in cultures where TS patients utter them. What is needed is more cross-cultural data before universal similarities and differences in cursing can be made with a degree of certainty.

Of course there are other bases for cross-cultural comparisons. Restricting the discussion to pathology, we can address one additional comparison based on 'culture-bound' syndromes which are not particular to the United States (Jay, 2000). I refer to conditions similar to people 'running amok', where victims of the nervous condition have licence to express intense emotions and even violent behaviour. Ataque de nervios, present in Latin America, and latah (in east Asia), are two such syndromes where women are given licence to express verbal and physical aggression, which would be stifled in public at other times.

A nervous condition or culture-specific syndrome provides an acceptable outlet for cursing not otherwise enjoyed in their culture and it would be interesting to record what kinds of speech occur during these episodes. At this point we see profanity is common across cultures on the basis of the outbursts of people who cannot control their cursing. It is also the case that religion plays a greater role in setting standards for public behaviour in some countries than others and it is to the issue of religious restrictions that we turn next.

Religion, learning and language restrictions

Social learning explains why we do or do not curse. For example, the child who is told that 'sex is bad and sinful' develops a negative emotional response to sexual stimuli. Also on the negative side, people have learned a negative emotional response to the word 'abortion' when told such things as 'abortion is murder'. The religiously raised child who has heard and read many positive emotional statements about God will be positively conditioned to this word. Religious parents forbid the use of profanity in the home (Jay, King and Duncan, 2004). But a child reared in a home filled with profanities will learn less positive reactions to religious words and concepts. Centuries of prohibitions on and declarations about the use of profanities restricted their use.

Religious people become conditioned to think of profanities as 'bad' words. Other words are 'good' words. 'Good' words are non-profane, non-obscene, those that do not offend or attack religion. 'Bad' words, like *goddamn, shit* or *fuck*, offend religious people, who will not utter them and do not want others to utter them. Religious training and practice creates social tensions within a culture about the behaviours and thoughts that must be inhibited. Tensions surrounding religion and religious figures at times require catharsis through humour and joking. Religious figures thus become the subjects of religious jokes. Legman (1975) recorded scores of jokes that include priests, nuns and other religious figures. He dedicated several pages to the subject of 'mocking God' in his second volume of jokes.

Restrictions on words originate in part from religious ceremonies and sacred texts (Bible, Koran). Words are defined as 'bad' through religious doctrine, Old Testament law, Islamic or other religious laws, or when religious authorities declare words and thoughts as forbidden. Religious ceremonies employ special language that is regarded more highly than everyday speech. Generally speaking, religious restrictions are based on the notion that words are 'good' or 'bad' and that 'bad' people use 'bad' words. One's attitude about religion and blasphemy depends on one's personal-psychological development and indoctrination in a religious community.

Censorship

Another means of teaching people that profanity is taboo is through the process of censorship. Words have to be sacred, powerful or dangerous to be censored by religions. An example of religious censorship over speech comes from the motion picture industry. From the first days of 'talking' pictures, the Catholic Church played a significant role in censoring American films (Jay, 1992, ch. 6). In 1927 a set of guidelines for film language, known as the 'Don'ts and Be Carefuls', banned '*god, lord, Jesus, Christ, hell, damn, gawd*, and every other profane and vulgar expression however it may be spelled' (Jay, 1992, p. 217). Here the public is explicitly informed that profanity is powerful through censorship standards.

The Church banned profanity because it had the power to do so. Recently, however, these prohibitions on profanity have declined significantly. Profanity is now common in all forms of popular media (radio, television, newspapers, comic strips). As older prohibitions on profanity have largely disappeared, current media censorship focuses on obscene and indecent speech (Flexner, 1976; Jay, 2000). Punishment and sanctions must be understood in light of definitions of profanity and blasphemy.

Profanity and blasphemy

To be *profane* means to be secular or indifferent toward religion. A profane word is not an attack on the Church; it amounts to indifference toward or a misuse of religious terminology. *Holy shit!* is a profanity. *Blasphemy* is more troublesome; it is an attack on religion and religious figures. It represents an intentional and offensive threat to religion subject to greater punishment than profanity (Jay, 1992, 1996). *The Pope is a fool* is a blasphemous statement. The distinction is necessary here even though the person on the street uses 'profanity' to refer to all categories of offensive speech. However, profanity and blasphemy are specific categories of religious speech sanctioned by religious authorities.

Censorship is enforced by members of a religious group or by one's (religious) parents. But if religious sanctions disappear in a community, profanities are frequently heard. In cultures where religion is powerful and its followers devout, penalties are proscribed which reduce the frequency of profanity and blasphemy. Since the 1900s in the United States, blasphemy prosecutions have all but disappeared. In comparison, Islamic punishments for blasphemy ('words of infidelity') still result in the loss of legal rights, marriage validation or claims to property (Elaide, 1987). However, there are ways for emotional expressions to sidestep the profane.

Euphemisms

One way around the religious restrictions on profanity is to express emotions through the use of euphemisms or substitute words. Euphemisms are milder replacement words (e.g., *cripes*) for more offensive counterparts (*Christ*!). The list includes expressions such as *darn, gosh darn, jeepers, heck, sugar, fudge* and *friggen*. K. C. Ushijima (2004) has documented how extensively Mormon students enrolled at BYU-Hawaii and on BYU campuses in Utah and Idaho use euphemisms and substitute words. As a testament to religious conservatism, Ushijima found that the most commonly used word by students on these three campuses was *crap*, which clearly contrasts with college students elsewhere who liberally utter words such as *fuck* and *shit* (Jay, 2000).

Religion is the source of some Americans' most frequent curse words because profanity *(damn)* is less offensive than sexually explicit (*cunt*) or aggressive speech (*fuck you*). Profanities (*damn, hell, Christ*) are acceptable in public speaking and in popular media in many cultures. Because profanity is so common and frequent in the United States, it is quickly learned by children, who along with others realize that they will not be punished as much for uttering profane epithets as they will for obscenities.

But people who define themselves as religious should eschew profanity regardless of how mild others deem it.

The religious personality and cursing

Personality refers to an individual's consistent patterns of behaviour and thought; examples are extraversion or neuroticism. We tend to think of personality as fairly stable across contexts, but we must realize that environment and learning also influence personality. The notion of personality allows us to differentiate individuals on the basis of personality traits and ask questions such as 'What kind of woman curses in public?'.

The answer to the question must reference personality factors because an act of cursing springs from a speaker's personality and speech habits. How seriously anyone treats profanity depends chiefly on her view of God. Cursing, as a habit, is part of a woman's psychological make up. When we hear a woman cursing we see traits related to her religiosity, aggressiveness, anxiety, racism or hostility. By training, a religious woman is more likely to be offended by profanity than a woman who is not religious.

Offendedness and offensiveness

Before we continue we need to distinguish between properties of words and properties of people. The notion of *offendedness* refers to a speaker's sensitivity to offensive language. Offendedness is an aspect of personality; it is a psychological reaction to words. In contrast, *offensiveness* is a property of words. Words can be very offensive or inoffensive. Offendedness is not innate, people learn to be offended by 'offensive' words. One's offendedness is a product of personality development and social awareness, which ultimately affect one's reaction to profanity and one's tendency to curse. A religious woman who is offended by profanity will not utter profanity. Similarly, a mother with high sex anxiety is unlikely to use sexual slang around her children.

To correlate cursing habits with personality, psychologists administer personality tests to subjects and then they measure their reactions to taboo words. Personality scores, for example, high sex anxiety (or religiosity) are then correlated with the word data. Very little work has been done to develop a test of offendedness and as a result we know little about what kind of women curse in public. While we have established facts about offensiveness (Jay, 1992), we know less about offendedness. Below is what one can find in the social science literature.

Long and Herrmann (1997) developed a 45-item questionnaire to gauge a person's sensitivity to taboo words and behaviours. Questions were designed to ask how acceptable, on a one-to-seven scale, respondents find behaviours related to questions tapping Sexuality, Religion, Obscenity, Seaminess, Liberality (live-and-let-live), Publicity (public displays of questionable behaviour) and Laxity (society has declined morally). Three of these factors (Publicity, Liberality and Laxity) provided consistent scores. The Publicity, Liberality and Laxity sub-scales measure feelings related to the sacred, which are used to make word choices in social situations. Gender differences were not delineated in the paper. This is just a beginning and more research is needed to fully develop the questionnaire with a non-college sample.

A review of personality and language research indicates that reliable correlations exist between cursing and religiosity and between cursing and sexual attitude (Jay, 2000). People with high religiosity and those with high sexual anxiety tend to be offended by profanity and sexual slang. These variables have also been examined with two additional lines of research, studies of viewers' reactions to speech on television and laboratory studies examining subjects' reactions to offensive speech.

Broadcast language

A good predictor of one's offendedness (by crude language) is the depth of religious belief. Complaints about television broadcast content are linked to religiosity and sexual conservatism. Here we learn little about gender. Hargrave (1991) and Wober (1980, 1990) recorded complaints about broadcast language and the complainers' demographics. Hargrave identified five groups of people with unique approaches to broadcast content:

1. 'Anti-sexual', who are mainly young men who were offended by sexual terms.
2. The 'offended', who are frequent churchgoers with strong and negative opinions about all types of offensive words on television.
3. 'Non-anatomical', who are most offended by scatological references and those words that referred to the genitals.
4. 'Permissive respondents', who are least likely to complain about sexual words.
5. 'Religious protectors', who are conservative churchgoers who reacted most strongly to words from religious origin.

These results are interesting in light of Long and Herrmann's (1997) work. Consistent predictors of reactions to speech in both studies were attitudes about the sacred, moral decline and public displays of offensive behaviour.

Laboratory studies

Religiosity has been an accurate indicator of one's hesitation to say taboo words in experimental settings. Here is better evidence about speaker gender. The explanation behind one's hesitation is as follows, when a speaker takes longer to say a taboo word relative to a neutral word, this hesitation represents the process of repression. Repression delays both decision processes and utterance latencies (reaction time, RT).

Grosser and Laczek (1963; see also Grosser and Walsh, 1966) compared students from parochial school backgrounds with students from secular schools to see if reluctance to say taboo words (utterance latencies) was related to religious training. Subjects viewed single words projected on a screen. They had to pronounce the word on the screen for the experimenter. The time between the end of the visual presentation and the onset of the oral report was recorded (RT). Subjects saw 15 neutral words, 15 aggressive words, 15 taboo sex words, then 15 more neutral words. The taboo sex words were (*prostitute, sperm, homosexual, pervert, adultery, douche, intercourse, erection, lesbian, seduce, vagina, penis, masturbation, rape* and *incest*). The RTs to the taboo words were significantly slower than any other sets of words, indicating that word meanings caused different reactions across participants.

The religiosity effect was most pronounced in the parochial females. The non-parochial females had the fastest RTs to the taboo words and the parochial females had the slowest RTs. The males fell between these extremes; the male secular subjects had the same RTs as the male parochial subjects for the taboo words. The authors attribute the parochial females' strong response repression effect to their moral training in school. Religious background and religious belief (some 40 years ago) have significantly affected measures of offendedness. One other potent variable in personality research is sexual repression.

Women's sexual anxiety, guilt, repression and cursing

Historically, religions have placed severe punishments on sexual expression (see Grey, 1993) and as a consequence people who are highly religious are often highly anxious about, and offended by, sexual language. Religiosity comingles with one's sexual anxiety. Hargrave (1991) found this in his broadcast speech survey, and numerous laboratory studies have confirmed the relationship between sex anxiety and repression.

Free-association research

A traditional approach to studying word meaning is the free association method. Subjects are presented with a target word and asked to respond with the first word that comes to mind. Galbraith, Hahn and Leiberman (1968) used the word-association test to examine the relationship between sex guilt and responses to *double entendre* words, which possessed substantial sexual connotation (*mount, pussy, screw*).

Subjects first completed the Mosher Forced-Choice Guilt Scale to measure their level of sexual guilt. Next their responses were recorded to a set of 50 words, 30 *double entendres* and 20 words devoid of sexual meaning. Associative responses to the words were scored 0, 1 or 2 depending on the amount of symbolic sexual components in the response. The higher the numerical score the higher the verbal, sexual response. Results indicated that sexual guilt was negatively correlated ($r = -0.41$) with sexual responsivity. Scores reflecting the frequency and flagrancy of verbal sexual responses in the free-association task to *double entendre* sex-slang terms were negatively correlated with guilt over sexuality.

The free-association format has been used to test sexual responsitivity in relation to males' repression and defensiveness. Schill, Emanuel, Pederson, Schneider and Wachowiak (1970) used free association to examine sexual responsivity to *double entendres* with a group of male college students. They found that personality traits of Defensiveness and Sensitization were related to the sexual responses provided during a free-association task with *double entendre* words with sexual connotations (*pussy, screw*). Subjects rated low in Defensiveness had the highest level of sexual responsitivity. Non-defensive Repressors and Sensitizers had greater sexual responsitivity than did Defensive Repressors. These results were obtained when male subjects were tested by a male experimenter.

When male subjects are tested by a female experimenter, sexual responsitivity is reduced. Under these circumstances male subjects become inhibited, because they want to make a good impression on the female experimenter and therefore they inhibit their sexual responses. The subjects' need to repress sexual responses is more salient with the female experimenter. With the male experimenter, male subjects are less defensive and more responsive without worrying about the impressions that their sexual responses makes on him.

Milner and Moses (1972) used both female and male subjects to extend the findings of Schill, Emanuel, Pederson et al. (1970). Using sexual responsivity measures to *double entendres* with both male and female experimenters, Milner and Moses found no overall differences comparing males' and females' responsivity. However, sexual responsiveness of males was significantly inhibited when the test was administered by a female experimenter. The sexual responsivity of the females tested by a male experimenter was significantly lower than all of the other experimental groups. Therefore, sexual associations to *double entendre* words were repressed when a member of the opposite sex administered the test and this was especially true for females. Experiments like these reveal the sexual dynamics underlying repression, which fit with those that underlie religiosity. However, the link between religiosity and gender is inferred and not empirically established by the researchers.

Women, power and taboo language

Gender identity is a set of beliefs, behaviours and norms that permeate human activity. Each culture seeks to transform infants into masculine and feminine adults. Gender identity is a set of cultural prescriptions and expectations that specify how men and women, gays and lesbians, should behave. In the past, cursing and aggression have been most closely identified with masculinity. Our cultures constrain *how* speakers communicate about sexuality. Sexuality is a taboo topic in the United States and words denoting sexual activity are avoided. Sexual speech is taboo because sexuality is taboo, not vice versa. Historically, American women have been expected to repress sexual thoughts, while men have been freer to use sexual speech.

Speaking sexually in public is intimately bound to cultural definitions of human sexuality and gender identity. A speaker's gender identity affects the tendency to curse in cultural contexts. Gender identity (with age, wealth, occupation and class) is a co-variant of power. The freedom to curse without punishment is for those who have power. But cursing and dominance are masculine traits, and ultimately cursing in public depends on both gender identity and power. Males tend to have more power to curse in the United States than females, though this is not universally true as we see later.

Throughout history men and women have experienced different standards for public behaviour. Not long ago, men cursed freely in pubic, especially in male-centered contexts, such as factories, taverns and sporting events. As women entered contexts historically occupied by males, women's cursing standards became more relaxed. Even though we have not experienced parity, American women can curse more openly in public now and men can no longer use obscenity as freely as in the past.

Research on gender and cursing reveals three recurrent findings, men curse more than women; men use a larger vocabulary of curse words than women; and men use more offensive curse words than women (Jay, 1992, 2000). Gender differences in cursing emerge when children enter school and they persist into middle age. It is worth noting, however, that women outnumber and outswear men in nursing-home settings (Jay, 1996). While men generally curse more in public than women, research indicates that the frequency gap between men's and women's swearing is decreasing (Jay, 2000). Gender differences in cursing are also related to differences in the use of sexual terms, joke telling, harassing speech, insulting and fighting words.

Sexual terminology

Gender differences regarding the use of sexual terminology have been documented many times. Heterosexual men, women, gay men and lesbians speak with distinctive sexual lexicons and prefer different terms for genitalia and sexual acts (Walsh and Leonard, 1974; Sanders, 1978; Sanders and Robinson, 1979; Terry, 1983, 1994; Wells, 1989, 1990). Men and women also write different kinds of sexual graffiti (Bruner and Kelso, 1980; Arluke, Kutakoff and Levin, 1987); that is, men's graffiti is more sexually suggestive and less socially acceptable relative to women's. Men's graffiti also tends to be more racist, more homophobic and less romantic than women's graffiti.

Joke telling

Speaker gender plays a significant role in dirty joke telling (Mitchell, 1985). Reliable differences appear in the selection of joke themes, characters in jokes, and forms of jokes. Men, relative to women, tell a higher percentage of obscene jokes, religious jokes, ethnic-racial jokes, and jokes about death and drinking. Women, relative to men, tell a higher percentage of absurd jokes, morbid jokes, Pollack jokes, jokes about authority figures and jokes with plays on words. Men tell more openly aggressive and hostile jokes than women. Finally, women prefer to tell their jokes to other women, while men are more willing to tell jokes to mixed audiences and opposite-sex audiences.

Harassment and fighting words

An interesting pattern of gender differences emerge when research on sexual harassment and fighting words is examined. Women are more sensitive and men less sensitive to speech that constitutes verbal sexual harassment (Jay and Richard, 1995). In contrast to sexual harassment dynamics, men are more sensitive to what constitutes fighting words (Jay, 1990). Fighting words are personally provocative words that lead to violence. Men are more likely than women to say they would be provoked into fighting by insulting or threatening speech. Women are more sensitive to harassment and men seem more sensitive to the dynamics of fighting language.

Gender-related insults

As for the question of words and insults, gender identity provides a basis for insulting words. Insults are based on cultural differences in men's and women's personalities. To get a clear picture of how men and women insult each other, one must first appreciate the kinds of traits associated with American men and women. Masculinity is associated with traits such as aggressiveness or dominance. Femininity is associated with traits such as nurturance and sensitivity. Gender-related insults tend to be based on references about deviations from expected or idealized gender-related behaviour.

Risch (1987) asked women to list insults for men and found that the most frequent words were based on references to the genitalia (*dick*), buttocks (*ass, asshole*) and ancestry (*bastard, son of a bitch*). Preston and Stanley (1987) asked subjects to list the 'worst thing' men and women could say to each other. They found the worst insults were:

woman to man: *bastard, prick*
man to woman: *cunt, slut*
man to man: *faggot, gay*
woman to woman: *bitch, slut*

The semantics of insult in these studies seem clear. Insults directed to heterosexual men refer to them as insincere or effeminate. Insults directed to heterosexual women refer to sexual looseness. These gender-related insults for women and men have legal implications as the dimensions of sexual looseness (*whore)* and homosexuality (*faggot*) are likely to be perceived as fighting words (Jay, 1990).

Insults are not merely offensive words; they are references to behaviours and traits that disturb Americans. The semantic structure of insults provides a model of those behaviours and traits. Through the use of detailed interviews with college students, Holland and Skinner (1987) constructed a cultural model of insulting. The semantic dimensions used in the model of insults were based on sexuality, attractiveness and sensitivity.

Several categories of terms were specific to males and females. Female directed insults were references to:

women who promised intimacy but did not fulfill the promise – *dickteaser*
social deviants who want too much from men – *bitch*
ugly, unattractive women – *scag, dog*
sexually loose women – *cunt, slut*

Insults for men also had definable target behaviours/traits:

effeminate or weak – *homo, fag, wimp*
insincere or mean – *bastard, prick, asshole*
inept, unattractive – *nerd, jerk*
attractive but sexually exploitative – *wolf, macho, stud*

Holland and Skinner (1987) showed that gender-related insults go beyond sexuality as a basis for insulting, as insults also reference attractiveness, ability, sexual potential and ineptitude. These dimensions of gender-based insults can be found in popular US media.

Media stereotypes

The construction of gender and gender-related insults is influenced by, and reflected in, media. American stereotypes are reinforced in the electronic and print media, as has been demonstrated in motion pictures (Jay, 1992), newspaper comic strips (Brabant, 1976; Brabant and Mooney, 1986; Mooney and Brabant, 1987, 1990; Jay, 1992) and televised films (Jay, 1993). The overwhelming majority of the portrayals of men and women cursing show that men curse more than women, men use more offensive words than women, women use more euphemisms than men. Men are rarely sanctioned for cursing. Women who curse tend to represent 'bad' characters (whores, drunks, drug users). The role of these exaggerated stereotypes of men and women are important to the degree to which they affect consumers.

A caution

One note of caution must be addressed on the issue of gender differences and speech. Henley (1995), reviewing literature on communication and dominance, concluded that women of colour are generally ignored in these studies, limiting applications to predominantly white middle-class society. Gender in most studies refers to white men and white women. Also, heterosexual identity is assumed in many gender studies. Analyses based on gay men's and lesbians' speech are less common in the literature. Obviously a broader sample of ethnic, homosexual and lower-economic groups is needed to draw valid conclusions about gender differences and cursing.

One of the most influential pragmatic forces controlling cursing is the power relationship between the speaker and the listener. Power is the ability to influence others through control over desired resources. Power co-varies with age, education, wealth, occupation, gender and race. Among equals, speakers adopt a level of verbal and non-verbal communication that is responsive to the listeners' power. Speaker–listener communication includes eye contact, personal space, speech volume, vocabulary, syntax and profanity. These components will shift according to the level of formality adopted. Power makes communication among non-equals asymmetrical. People with power have licence to tell jokes, make fun of subordinates and use curse words. The level of speech formality adopted in a context depends on who has the power to shift levels up or down.

Cursing generally occurs at an informal, non-standard level. Cursing should be appropriate for the speaker–listener relationship. Speakers can 'talk up' or 'talk down' to the listener, urging a shift to higher or lower standards of formality. A speaker can initiate the use of cursing as a way to move to a non-standard, more relaxed level of speaking. A working-class woman, however, might inhibit cursing when she thinks she might be judged negatively by her boss.

Working-class women and cursing

Paul Fussell (1983) described many of the obvious and not so obvious differences in American lifestyles as a function of status. One lifestyle difference is our speech patterns. He noted that the 'bohemian class' is fairly free to use obscene speech, using it with rhetorical effectiveness. Working-class speakers are fonder than most people of calling someone an *asshole* according to his analysis. Fussell stated, 'your social class is still most visible when you say things' (p. 151), noting that the sizeable middle class feared offending others. To avoid offence they employ euphemisms, genteelism and mock profanity ('holy cow'). Examining both gender and class differences, Hughes (1992) noted the reluctance of lower-working-class women to use profanities at work. In contrast to their lack of profanity, the lower-working-class women frequently use expletives, in part to maintain social cohesion.

According to Hughes (1992), the use of 'prestigious' standard English has little value for lower-working-class women because standard English cannot provide any social advantage to them or increase their chances for success. Speaking standard English would work to isolate them from their own peers. In contrast to the stereotype that women curse less than men, lower-working-class women frequently use strong expletives (*cunt, fuck, shit*) that many middle-class men avoid. Daly, Holmes, Newton and Stubbe (2004) drew a similar conclusion about the use of expletives on the factory floor. Many exchanges using the word *fuck* served to promote cohesion and solidarity between women (and men) factory workers. Interestingly, working-class women in Hughes's sample exhibit a strong moral code against the use of profanity (*Jesus, Christ, God*). The middle class might find these values surprising, that is, where obscenities are acceptable and even encouraged, but profanities are avoided. One cannot ignore the impact of religion on this choice.

In addition there are some data relating joke telling to occupational status. Coser (1960), studying joking among staff members of a mental hospital, found that the most frequent targets of the senior staff joking were junior staff members. Patients and relatives were targeted by the junior staff members. Humour was directed downward at those with no power over the speaker.

Jokes and harassment at work

In many occupational settings, speaker power is a defining feature of sexual harassment. Most verbal sexual harassment suits involve junior female workers claiming to have been harassed by male managers. Unwanted jokes, obscenity, sexual innuendo, comments about physical attractiveness or appearance flow from the top of the hierarchy down. Top-down harassment has been documented with nurses and doctors (Cox, 1987, 1991a, 1991b; Braun, Christle, Walker and Tiwanak, 1991), medical students and physicians (Nora Daugherty, Hersh et al., 1993), workers and management (Martell and Sullivan, 1994).

Henley (1995) studied non-verbal communication patterns as a function of power, gender and dominance. Her comprehensive view of communication presents cursing within a broad interpersonal context which incorporates class, race and gender variables. Power gives a speaker the licence to do things that the powerless cannot. Dominance legitimizes invasions of personal space, touching others, engaging in eye contact, and addressing subordinates by their personal names rather than by title. Dominance and power also legitimize the use of offensive language over subordinates. Therefore, the doctor tells a dirty joke and the nurses laugh, but not vice versa.

Conclusion

What have we learned by examining the nexus of gender, cursing and religion? We see that each factor has a significant impact on language choices. Women curse less than men, generally speaking, but times are changing and for working-class women this generalization does not apply. Social class, status, power and occupation are important mitigating variables. The Church has lost power to broadly censor speech in US media; however, the faithful maintain values that allow them to repress profanity in public. Religiosity and sexual anxiety are primary traits underlying language repression. Cross-cultural studies reveal the pervasiveness of cursing and the semantics of the forbidden in different cultures; some more focused on profanity than others. Finally, the role of cursing for American women is changing and we will probably continue to see a relaxation of the restriction on their cursing in the future.

References

Arluke, A., Kutakoff, L., and Levin, J. (1987). Are times changing? An analysis of gender differences in sexual graffiti. *Sex Roles* 16, 1–7.

Brabant, S. (1976). Sex role stereotyping in the Sunday comics. *Sex Role* 2(4), 331–7.

Brabant, S., and Mooney, L. (1986). Sex role stereotyping in the Sunday comics: Ten years later. *Sex Roles* 14(3/4),141–8.

Braun, K., Christle, D., Walker, D., and Tiwanak, G. (1991). Verbal abuse of nurses and non-nurses. *Nursing Management* 22, 72–6.

Bruner, E., and Kelso, J. (1980). Gender differences in graffiti: A semiotic perspective. *Women's Studies International Quarterly* 3, 239–52.

Cardoso, F., Veado, C., and de Oliveria, J. (1996). A Brazilian cohort with Tourette's syndrome. *Journal of Neurology, Neurosurgery, and Psychiatry* 60, 209–12.

Coser, R. (1960). Laughter among colleagues: A study of the social functions of humour among staff of a mental hospital. *Psychiatry* 23, 81–95.

Cox, H. (1987). Verbal abuse in nursing: Report of a study. *Nursing Management* 18, 47–50.

Cox, H. (1991a). Verbal abuse nationwide, Part I: Oppressed group behavior. *Nursing Management* 22, 32–5.

Cox, H. (1991b). Verbal abuse nationwide, Part II: Impact and modifications. *Nursing Management* 22, 66–9.

Crystal, D. (1987). *The encyclopedia of language*. New York: Cambridge University Press.

Daly, N., Holmes, J., Newton, J., and Stubbe, M. (2004). Expletives as solidarity signals in FTAs on the factory floor. *Journal of Pragmatics* 36, 945–64.

Elaide, M. (Ed). (1987). *The encyclopedia of religion*. Vol. 2. New York: Macmillan.

Flexner, S. (1976). *I hear America talking*. New York: Van Nostrand.

Fussell, P. (1983). *Class: A guide through the American status system*. New York: Summit.

Galbraith, G., Hahn, K., and Leiberman, H. (1968). Personality correlates of free-associative sex responses to double-entendre words. *Journal of Consulting and Clinical Psychology* 32, 193–7.

Grey, A. (1993). *Speaking of sex: The limitations of language*. London: Cassell.

Grosser, G., and Laczek, W. (1963). Prior parochial vs secular secondary education and utterance latencies to taboo words. *The Journal of Psychology* 55, 263–77.

Grosser, G., and Walsh, A. (1966). Sex differences in the differential recall of taboo and neutral words. *The Journal of Psychology* 63, 219–27.

Hargrave, A. (1991). *A matter of manners: The limits of broadcast language*. London: John Libbey.

Henley, N. (1995). Body politics: What do we know today? In: P. Kalbfleisch and M. Cody (Eds), *Gender, power, and communication in human relationships*. Hillsdale, NJ: Erlbaum (pp. 27–61).

Holland, D., and Skinner, D. (1987). Prestige and intimacy. In: D. Holland and N. Quinn (Eds), *Cultural models in language and thought*. New York: Cambridge University Press (pp. 78–111).

Hughes, G. (1991). *Swearing*. Oxford: Basil Blackwell.

Hughes, S. E. (1992). Expletives of lower working-class women. *Language in Society* 21, 291–303.

Jay, T. B. (1980). Sex roles and dirty word usage: A review of the literature and reply to Haas. *Psychological Bulletin* 88, 614–21.

Jay, T. B. (1990, March). What are 'fighting words'? Presented at Eastern Psychological Association Meeting, Philadelphia.

Jay, T. B. (1992). *Cursing in America*. Philadelphia: John Benjamins.

Jay, T. B. (1993, April). Cursing: Too taboo for television. Paper presented at Popular Culture Association Meeting, New Orleans.

Jay, T. B. (1996). Cursing: A damned persistent lexicon. In: D. Herrmann et al. (Eds), *Basic and applied memory research: Practical applications*, Vol. 2. Mahwah, NJ: Erlbaum (pp. 301–13).

Jay, T. B. (2000). *Why we curse*. Philadelphia: John Benjamins.

Jay, T. B., King, K., and Duncan, T. (2004). *Soap in my mouth: College students' narratives and attitudes about punishment for cursing*. Paper presented at a meeting of the Eastern Psychological Association, Washington, DC.

Jay, T. B., and Richard, D. (1995, April). Verbal sexual harassment, figurative language, and gender. Presented at Eastern Psychological Association Meeting, Boston.

Lees, A. (1985). *Tics and related disorders*. Edinburgh: Churchill Livingstone.

Lees, A., and Toloso, E. (1988). Tics. In: J. Jankovic and E. Tolosa (Eds), *Parkinson's disease and movement disorders*. Baltimore: Urban and Schwarzenberg (pp. 275–81).

Legman, G. (1975). *No laughing matter: Rationale of the dirty joke* (second series). New York: Bell.

Lieh-Mak, F. L., Chung, S. Y., Lee, P., and Chen, S. (1982). Tourette Syndrome in the Chinese: A follow-up of 15 cases. In: A. Friedhoff and T. Chase (Eds), *Gilles de la Tourette Syndrome*. New York: Raven Press (pp. 281–3).

Long, R., and Herrmann, D. (1997, April). Personality and cursing. Presented at Eastern Psychological Association Meeting, Washington, DC.

Martell, K., and Sullivan, G. (1994). Sexual harassment: The continuing workplace crisis. *Labor Law Journal* 45, 195–207.

Milner, J., and Moses, T. (1972). Sexual responsivity as a function of test administrator's gender. *Journal of Consulting and Clinical Psychology* 39, 515.

Mitchell, C. (1985). Some differences in male and female joke-telling. In: R. Jordan and S. J. Kalcik (Eds), *Women's folklore, women's culture*. Philadelphia: University of Pennsylvania press (pp. 163–86).

Mooney, L. A., and Brabant, S. (1987). Two martinis and a rested woman: 'Liberation' in the Sunday comic. *Sex Roles* 17(7/8), 409–20.

Mooney, L. A., and Brabant, S. (1990). The portrayal of boys and girls in six nationally-syndicated comic strips. *Sociology and Social Research* 74(2), 118–26.

Nomura, Y., and Segawa, M. (1982). Tourette Syndrome in Oriental children: Clinical and pathophysiological considerations. In: A. J. Friedhoff and T. N. Chase (Eds), *Gilles de la Tourette Syndrome*. New York: Raven Press. (pp. 277–80).

Nora, L. M., Daugherty, S., Hersh, K., Schmidt, J., and Goodman, L. J. (1993). What do medical students mean when they say 'sexual harassment'? *Academic Medicine* 68(10), 49–51.

Preston, K., and Stanley, K. (1987). 'What's the worst thing ... ?' Gender-directed insults. *Sex Roles* 17, 209–18.

Regeur, L., Pakkenberg, B., Fog, R., and Pakkenberg, H. (1986). Clinical features and long-term treatment with pimozide in 65 patients with Gilles de la Tourette syndrome. *Journal of Neurology, Neurosurgery, and Psychiatry* 49, 791–5.

Risch, B. (1987). Women's derogatory terms for men: That's right, 'dirty' words. *Language in Society* 16, 353–8.

Robertson, M., and Trimble, M. (1991). Gilles de la Tourette syndrome in the Middle East: Report of a cohort and a multiply affected large pedigree. *British Journal of Psychiatry* 158, 416–19.

Sanders, J. (1978). Male and female vocabularies for communicating with a sexual partner. *Journal of Sex Education and Therapy* 4, 15–18.

Sanders, J., and Robinson, W. (1979). Talking and not talking about sex: Male and female vocabularies. *Journal of Communications* 29, 22–30.

Schill, T., Emanuel, G., Pederson, V., Schneider, L., and Wachowiak, D. (1970). Sexual responsivity of defensive and nondefensive sensitizers and repressors. *Journal of Counseling and Clinical Psychology* 35, 44–7.

Shapiro, A., Shapiro, E., Young, J., and Feinberg, T. (1988). *Gilles de la Tourette syndrome* (2nd Edn). New York: Raven Press.

Terry, R. (1983). A connotative analysis of synonyms for sexual intercourse. *Maledicta* 7,237–53.

Terry, R. (1994). Synonyms for sexual intercourse: Evidence of Zipf's Law. *Psychological Reports* 75, 1669–70.

Ushijima, K. C. (2004). *Substitute swear words in the Mormon culture*. Senior Thesis. Brigham Young University-Hawaii.

Walsh, R., and Leonard, W. (1974). Usages of terms for sexual intercourse by men and women. *Archives of Sexual Behavior* 3, 373–6.

Wells, J. (1989). Sexual language usage in different interpersonal contexts: A comparison of gender and sexual orientation. *Archives of Sexual Behavior* 18, 127–43.

Wells, J. (1990). The sexual vocabularies of heterosexual and homosexual males and females for communicating erotically with a sexual partner. *Archives of Sexual Behavior* 19, 139–47.

Wober, J. (1990). Language and television. In: H. Giles and W. Robinson (Eds), *Handbook of language and social psychology*. New York: Wiley (pp. 561–82).

Part II

Gender and Language Use in Religious Communities

5
Women and Men: Languages and Religion in Taiwan

Chao-Chih Liao

The earthquake in Taiwan in September 1999 resulted in renewed searching within that country – an awakened awareness of the role of religion. The reasons the Taiwanese people have chosen to follow a particular religion are varied. Many young people in Taiwan are drawn to Christianity[1] because Christian organizations offer English classes whereas many older women and men remain loyal to Taoism and Buddhism. This chapter explores some connections between gender, language and religion in Taiwan and the reasons why educational institutions have contributed to the rise of Christianity in particular.

Most Taiwanese religion is a combination of Confucianism, Buddhism and Taoism. However, because Taoism allows for more materialistic pursuits, it is embraced by many of today's Taiwanese.[2] Those who follow the Taoist religion may be looked down upon by the more conservative Buddhists. I have heard it said that if a person's faith is expressed only by participation in the *chin-hsiang-tuan* ('Incense Holding Parade') during festivals, s/he is in an 'elementary-school' phase of religious development, while those who study the Buddhist scriptures would consider themselves to be in the 'university' phase. A Buddhist believes that human lives are miserable and it would be better not to be born. Once born, however, one must refrain from greed, anger and silliness, and pray not to return to life through reincarnation. Chen (1996) summarizes the complexities and contradictions by saying that the Taiwanese religious experience begins with ancestor homage, is cultivated by Confucianism, coloured by Taoism, and modified by Buddhism. That is to say, religion in Taiwan is complex and multi-layered.

From elementary school onward, files are kept for all Taiwanese, in which their religion is identified. Ten 50-year-old women interviewed for this study stated that their religion was Buddhism, although six of them admitted they were actually Taoists: they differentiated on issues like reading scriptures or strict adherence to a vegetarian diet. They, like many Taiwanese and Hakkas, prepare food and burn incense in front of the house, at the ancestor hall and/or at temples only on specific days and not on a regular basis. Such contradictions are common in today's Taiwan.

This study applied three research methods to explore gender, language and religion: participant observation, interview and questionnaire analysis. Participant observation and interviews lead to the following results: Taoist and Buddhist places of worship communicate in the Taiwanese or Mandarin languages in order to be believer-friendly, and Christian churches use English, Taiwanese, Mandarin, Hakka and aboriginal languages for this purpose. Buddhism uses the Mandarin transliteration of Sanskrit to reveal mysticism and tradition (many scriptures are Chinese transliterations from Sanskrit in the third to tenth centuries). Islam uses both Arabic and Mandarin.

The state of religion, gender and language classes in Taiwan

According to the Executive Yuan (1995) 52.8 per cent of all Taiwanese are religious believers (over 11 million people). Among believers, 43.7 per cent identify themselves as Buddhists, 34.7 per cent as Taoists, 8.5 per cent as I-kuan Taoists, 6.5 per cent as Christians, and 6.6 per cent as belonging to other religions.

In Taiwan, Buddhism and Taoism are the two main religions (86.9 per cent of the religious population identify themselves as Buddhist or Taoist); Christianity and Islam play more minor roles. The Taiwanese traditionally use incense to show respect to ancestors and in praying to ancestors for their help. Christianity and Islam do not allow this practice and consider it to be idol worship, which makes it difficult for the Taiwanese to give credence to these religions.

In the 1950s there were ten universities in Taiwan, three of which were Christian (Wang, 1996). Since the 1990s, many more universities, with new Departments of Religion, have been established. Their purpose is to promote the social and humane study of world religions, and thus to promote harmony among people of different faiths. Other steps have been taken as well, such as the Taiwan government's 2000–1 Show of Western Religions. The motivations for this exhibition were to bridge the gap between believers and non-believers and specifically to promote Christianity in Taiwan (Chen, 2000).

Though Islam is a major world religion, it is much less significant in Taiwan. There have been efforts to promote it, such as in 1990, when the ROC (Republic of China in Taiwan) government sold a piece of land at a very low price for the building of the Taichung Mosque. This was an attempt to please the international community, even though the ROC government clearly promotes Christianity over Islam in more significant ways.

One Buddhist master in Taiwan, Shih Zheng-yen, the founder of Tzu Chi, said, 'If one worships Buddha without studying scriptures, s/he becomes either superstitious or ignorant' (in Shih, 1992, p. 44). She views many Taiwanese as less devout since they do not read Buddhist scriptures or follow the main tenets of Buddhism. Because women tend to visit religious places and perform the rituals rather than study the scriptures, they are often viewed as more superstitious and ignorant than men.

The English word for Taoist and Buddhist places of worship is *temple*. However, in the Taiwanese language, the definitions are more complicated: Buddhists use the word *si* for a Buddhist temple, and other words (*miao, gong* or *tzu*) to identify Taoist places of worship. Many Taoists themselves are not aware of these differences. While collecting data, I found many university students to be unsure whether they were Taoists or Buddhists, identifying themselves as either Buddhist or Taoist, depending on the circumstances.

Religion and education

The *United Daily News* (26 March 2004, page A4) reported the Dean of Academic Affairs of the National Taiwan University as saying that NTU had informed its professors that they were not to initiate discussions about sex, politics or religion in class unless the course specifically focused on such material. While these three topics remain taboo in Taiwan's most prestigious university, newspapers and television newscasts continue to report about them and the students in this study cooperated well on the topic of religion for the purpose of data collection. However, religion is seldom the main topic of discussion in educational institutions. In a study of 1278 English teachers in Taiwan, only one male student noted that a female teacher was a pious Buddhist (Liao, 2005) and that was because she told her students about her religious activities. Such teachers can be reprimanded if students complain.

The link between Christianity and education in Taiwan is significant. Here, Christian institutes of higher education have a longer history than those of other religions. Chang-jong Catholic High School has existed since 1885. The first Catholic university in China/Taiwan is Fu-ren Catholic University, founded around 1930 and re-established in Taiwan in the 1950s. Since 1963, the Stella Matutina Girls' High School and Viator High School, two highly competitive schools in Taichung City, have been famous for quality education. Both are Christian institutions.

There are 25 Buddhist academic institutions, but only five Buddhist graduate schools. Temples offer courses to children on Confucius's teaching about filial piety and the hierarchical structure of society. These Buddhist institutions have been in existence only since 1994, more than 20 years after the founding of the Tzu Chi-initiated aid programmes for victims of natural disasters. Since Buddhism and Christianity each have their own higher education institutes in Taiwan, both systems can choose scholars from among the most brilliant students. Students attending the Christian institutions tend to be better off economically, partly because the Christian institutes of higher education are more established.

Among the smaller Taoist sub-sects, I-kuan Tao is the most enthusiastic about contributing to Taiwanese society, operating 34 nursery schools, 21 hospitals, 8 clinics and 3 nursing homes. Taoism has no educational institute. One reason Taoists have not established schools might be due to its tradition of personal devotion rather than promoting a doctrine of ambition and accomplishment.

In the late 1990s, the monks at the Chung Tai Chan Monastery proposed to cut the hair of all attendees at their summer camp and convert all of them to Buddhism. Many parents rigorously objected. The Taiwanese admire religion but despise such extremism and coercion. After the matter was settled, the Minister of Education, in hopes of developing both awareness and tolerance, proposed that Taiwanese universities offer courses on a variety of religions rather than focusing on one in particular. To most Taiwanese religion is suitable as a pastime, but religious believers who forsake their families to live and serve in temples or monasteries might be described as *tzou-huo-ru-mo* ('walking on the fire and stepping into devil'), referring to the rejection of family loyalty for the sake of another calling. Such abandonment of family is not encouraged in the culture at large, even for Buddhists.

Languages and religions in Taiwan

Taiwanese languages include the High Language of Mandarin, Low Taiwanese, the in-group Hakka and the many aboriginal languages (Huang, 1993). Mandarin speakers comprise at least 90 per cent of the population, Taiwanese speakers 78 per cent, and Hakka speakers 12 per cent (Huang, 1993). The Taiwan people are generally bilingual, some Mainlanders may be monolingual, and Hakkas and aborigines are often trilingual. A Taiwanese person who speaks English fluently enjoys high social status. Because of this, more Taiwanese women go to church, where there are times of English worship, and may eventually convert to Christianity. As such, women's language is often associated with higher status (Labov, 1972).

Christianity in Taiwan is a legacy of much missionary effort from the Western world. From 1624–62, missionaries from Holland learned the languages of Taiwan and worked to convert native tribes. The work was stopped in 1662 but recommenced in 1865 under the direction of the Presbyterian Church of England. In 1872, the Presbyterian Church of Canada started a Protestant mission in Taiwan (Campbell, 1889). In the 1900s and 1910s, Marjorie Landsborough (1972) served as a missionary for the Presbyterian Church of England in Formosa. Though Taiwan was ruled by Japan from 1895–1945, she taught classes in Taiwanese rather than Japanese, using Bible lessons to help boys and girls become literate.

Christianity has contributed much to Taiwan's languages. Missionaries helped compile the Hakka–English/English–Hakka dictionary in 1959 as well as the English–Taiwanese/Taiwanese–English dictionary, which used romanized spellings throughout. Taoist and Buddhist professionals have tried to appeal to visitors or new converts by using the local languages of Hakka or Taiwanese. Because aborigines reject Buddhism and Taoism, aboriginal languages are not used within those religions.

Islam first came to Taiwan in 1662 but many Muslims eventually converted to Taoism or Buddhism. It was not until after World War II that more Muslims arrived, this time with the support of the Chinese Nationalist Government. Most local Muslim men are older, born in 1949 or before.

Buddhism enjoys a higher status than Taoism. A real Buddhist tells people that s/he is a Buddhist and may condemn those Taoists who claim to be Buddhists. Most Taiwanese claim that Chinese believe in a combination of Confucianism, Taoism and Buddhism. However, they reject the idea of Confucianism as a religion, because Taoism and Buddhism have ritual ceremonies, while Confucianism does not – it is more a philosophy that values filial piety than a religion to be adhered to.

Taiwan's aboriginal groups have their own religion. However, many of them also believe in Christianity. From 1895–1945, the aboriginal people were isolated and no religious groups reached their settlements in the mountains to promote other religions. After the Japanese withdrawal in World War II, the Presbyterians arrived and converted 70 per cent of Taiwan's aborigines to Christianity (Yisikakafute, 1997).

The study

Results of interview and participant observation

When I questioned people whether they believed more men or women visit religious places of worship (temples and churches), the answer was overwhelmingly 'women'. Church visitors might be family members or younger people either single or not old enough to have children. However, visitors to Buddhist or Taoist temples are mostly older women (whose family members are either too busy or do not share the same religion); occasionally they attend with children and grandchildren. Women demonstrate gratitude and faith in their prayers – they pray for their husbands to succeed in their careers, their sons to do well in school, and their daughters to have good marriages. They also pray for good health for their families. Men tend to be more pragmatic in their religious observances – their prayers are more likely to be for material success.

Many Taiwanese students attend churches that offer English Bible and Conversation classes in order to improve their English. In Taiwan, one can find churches or temples using monolingual messages of Mandarin, Taiwanese or Hakka. However, it is impossible to find a Buddhist or Taoist temple using English in worship. In some churches, four languages – Mandarin, Taiwanese, English and Hakka/one aboriginal language – are used to cater to different groups of people at different sessions.

Questionnaire analysis

In March 2004, I distributed two sets of questionnaires to undergraduates; each participant completed only one set. For Set 1, their responses were immediate; Set 2 had two weeks to consider their responses. There were 32 responders in Set 1, 104 responders in Set 2.

Set 1

I asked a class of 32 university men and women to think of all the religious believers that they personally knew and to identify the gender, age, religion and the reason for conversion if they were aware of it. One hundred and thirty-nine believers – 93 female (66.9 per cent) and 46 male (33.1 per cent) – were listed. The numbers for each religion break down as follows:

Buddhist	56	(40.3%)
Christian	56	(40.3%)
Taoist	27	(19.4%)

Table 5.1 Ratio of older to younger people within their religions

	Buddhism	**Christianity**	**Taoism**	**Total**
Younger	24 (34.3%)	42 (60%)	4 (5.7%)	70
Older	32 (46.4%)	14 (20.3%)	23 (33.3%)	69
Total	56 (40.3%)	56 (40.3%)	27 (19.4%)	139

Chi-square value (χ^2) = 28.508 (p-value = 0.000).

The believers ranged in age from 15–95; the average age was 41.7 years. Because of this, I divided the list into two groups and set 42 years of age as the dividing line between the younger and older age groups. The gender variable seemed relatively evenly divided between the two groups; the age factor appeared to be more significant.

As can be seen in Table 5.1, more younger people believe in Christianity, while more older people believe in Taoism.

Set 2

For Set 2, each participant was given two weeks to interview two devout religious believers – one of their parents' age and the other closer to their own age – and to take note of the believers' age, gender, religion, motivation for their faith and the languages used in the place of worship they attended. The interview was then used for discussion in an English conversation and writing class. Four females out of the 104 total participants could not find any young believers and asked for permission to interview two older people. Six students visited one older believer only. It was generally observed that there were fewer younger believers to interview than there were older. This may in part be due to the fact that, when compared with the life experiences of younger people, older people have likely seen more illnesses, disasters, unpredictable events and hard-to-explain coincidences or miracles.

Seventy-five females and 29 males finished the assignment. Because men and women tend to interact with people of their own gender (Liao, 2005; Tannen, 1986), I expected the male participants to interview more male believers and the female students to interview more female believers. However, this was not the case. In both groups, more females than males were interviewed as believers.

The figures in Tables 5.2 and 5.3 suggest that the number of both older and younger Taiwanese female and male believers do not differ significantly within these religions. The gender variable seems quite balanced. However, Table 5.4 significantly demonstrates the age factor as salient here. Table 5.5 shows the ratio consistency in Sets 1 and 2 for the older groups, Table5.6 inconsistency in the younger groups:

Table 5.2 Older believers' religions

	Buddhism	**Christianity**	**Falungong**	**Taoism**	**Total**
Females	23 (29.1%)	16 (20.3%)	1 (1.3%)	39 (49.4%)	79
Males	11 (42.3%)	7 (26.9%)	0	8 (30.8%)	26
Total	34 (32.4%)	23 (21.9%)	1 (1%)	47 (44.8%)	105

$\chi^2 = 3.290$ (p-value = 0.349).

Table 5.3 Younger believers' religions

	Buddhism	**Christianity**	**Taoism**	**Total**
Females	17 (26.2%)	30 (46.1%)	18 (27.7%)	65
Males	3 (10%)	14 (46.7%)	13 (43.3%)	30
Total	20 (21.1%)	44 (46.3%)	31 (32.6%)	95

$\chi^2 = 4.084$ (p-value = 0.130).

Table 5.4 Older versus younger believers' religions

	Buddhism	**Christianity**	**Taoism**	**Others**	**Total**
Older	34 (32.4%)	23 (21.9%)	47 (44.8%)	1 (1%)	105
Younger	20 (21.1%)	44 (46.3%)	31 (32.6%)	0	95
Total	54 (27%)	67 (33.5%)	78 (39%)	1 (0.5%)	200

$\chi^2 = 14.029$ (p-value = 0.003).

Table 5.5 Comparison of Set 1 and Set 2 (older groups)

	Buddhism	**Christianity**	**Taoism**	**Others**	**Total**
Table 5.1	32 (46.4%)	14 (20.3%)	23 (33.3%)	0	69
Table 5.2	34 (32.4%)	23 (21.9%)	47 (44.8%)	1 (1%)	105
Total	66 (37.9%)	37 (21.3%)	70 (40.2%)	1 (0.6%)	174

$\chi^2 = 4.210$ (p-value = 0.240).

Table 5.6 Comparison of Set 1 and Set 2 (younger groups)

	Buddhism	**Christianity**	**Taoism**	**Others**	**Total**
Table 5.1	24 (34.3%)	42 (60%)	4 (5.7%)	0	70
Table 5.3	20 (21.1%)	44 (46.3%)	31 (32.6%)	1 (1.1%)	95
Total	44 (26.7%)	86 (52.1%)	35 (21.1%)	1	165

$\chi^2 = 17.861$ (p-value = 0.00).
(*Note*: the p-value of 0.00 in Table 5.6 is owing to my precision re: one's definition of Taoism. In response, some respondents changed their minds.)

Of the 105 older believers, only three female Taoists and one male Catholic had chosen a profession within their religion. The three older women operated *shentan* ('A shop selling specific gods for worship'[3]) at home and the male was a Catholic Father. All the remaining older believers donated money and labour to their religion. Of the 95 young believers, marginally more young males chose a religious career than did females – three Taoists (one female and two males) and one male Catholic. The female Taoist worked in a temple and the two male Taoists helped at their parents' *shentan*. The Catholic male attended a Catholic high school.

Languages used in places of worship

According to the reports about older believers (Table 5.7), 75 per cent of Buddhists communicate in Taiwanese, 61.1 per cent in Mandarin and 2.8 per cent in Hakka. The one place where Hakka was spoken was in a place where two other languages – Mandarin and Taiwanese – were also used. For Christians, 20 places of worship (87 per cent) communicated in Mandarin; seven places (30 per cent) used Taiwanese; four places (17 per cent) used English, and one place (4 per cent) used only the local aboriginal language. At 26 Taoist places of worship (55.3 per cent), Mandarin was spoken; at 28 places (59.6 per cent) Taiwanese and at one place both Taiwanese and Hakka were spoken. In summary, the main language of Buddhism and Taoism is Taiwanese, the main language of Christianity is Mandarin.

For younger people (Table 5.8), the main language used in Buddhist places of worship was Mandarin (14 places, 73.7 per cent); in Christian

Table 5.7 Languages used at the gathering places for older believers

	A	HT	M	ME	MHT	MS	M and T	T	Total
Buddhism	0	0	8	0	1	1	12	14	36
Christianity	1	0	11	4	0	0	5	2	23
Taoist	0	1	9	0	0	0	17	20	47

A: Aboriginal language; E: English; H: Hakka; M: Mandarin; S: Sanskrit; T: Taiwanese.

Table 5.8 Language used at the gathering places for young believers

	E	H	M	ME	MT	MTE	T	TS	Total
Buddhism	0	0	8	0	6	0	4	1	19
Christianity	1	0	23	11	5	2	2	0	44
Taoism	0	1	4	0	10	0	14	0	29

places, Mandarin (41 places, 93.2 per cent); and in Taoist places, Taiwanese (24, 82.8 per cent). The reason for the majority of older Buddhists speaking Taiwanese and the majority of younger Buddhists speaking Mandarin may not be that they attend different temples but that they simply did not pay attention to the languages they were not familiar with. If young people are fluent in Mandarin, they may ignore the fact that the Buddhist temple they attend also uses Taiwanese. One Catholic reported that only English was used in her church, because she only attended the English mass.

Motivations

This section concentrates on the believers' motivations for their faith. The pragmatic reasons for becoming religious may be personal need, concern for family, or out of gratitude. Examples of the first include wanting peace of mind or help to find a good spouse or a good job. Instances of the second might be to pray for healing for a family member or because it is the family religion. An example of gratitude could be when one's life is miraculously spared, as for those who survived the earthquake of 1999. Tables 5.9, 5.10 and 5.13 demonstrate common motivations for all Taiwanese, and Table 5.11 indicates that 12 older women are religious out of gratitude. Of these 12, two became Buddhists after the earthquake, nine became Taoists for different reasons, and one became a Catholic because the church helped her after the earthquake.

Table 5.9 Motivations for religious belief of older people

	F	FG	Fr	G	O	S	U	Total
Buddhism	11	0	3	2	2	14	4	36
Christianity	8	0	0	1	1	11	2	23
Falungong	0	0	0	0	0	1	0	1
Taoism	14	1	1	8	1	18	5	48
Total	33 (30.6%)	1 (0.9%)	4 (3.7%)	11 (10.2%)	4 (3.7%)	44 (40.7%)	11 (10.2%)	108

χ^2 = 10.862 (p-value = 0.900); F: family; G: gratitude; Fr: friends; O: others; S: self; U: unknown.

Table 5.10 Motivations for religious belief of younger people

	F	FG	Fr	G	O	S	U	Total
Buddhism	8 (38.1%)	0	0	2 (9.5%)	1 (4.8%)	9 (42.9%)	1 (4.8%)	21
Christianity	17 (38.6%)	1 (2.3%)	0	2 (4.6%)	4 (9.1%)	14 (31.8%)	6 (13.6%)	44
Taoism	18 (58.1%)	0	1 (3.2%)	2 (6.5%)	0	8 (25.8%)	2 (6.5%)	31
Total	43 (44.8%)	1 (1.0%)	1 (1.0%)	6 (6.3%)	5 (5.2%)	31 (32.3%)	9 (9.4%)	96

χ^2 = 11.287 (p-value = 0.504).

Table 5.11 Older men and women's motivations

	F	FG	Fr	G	O	S	U	Total
Women	26 (32.9%)	1 (1.3%)	2 (2.5%)	11 (13.9%)	3 (3.8%)	28 (35.4%)	8 (10.1%)	79
Men	6 (23.1%)	0	2 (7.7%)	0	1 (3.9%)	14 (53.9%)	3 (11.5%)	26
Total	32 (30.5%)	1 (1%)	4 (3.8%)	11 (10.5%)	4 (3.8%)	14 (53.9%)	3 (11.5%)	105

$\chi^2 = 7.631$ (p-value = 0.266).

Table 5.12 Younger men and women's motivations

	F	FG	Fr	G	O	S	U	Total
Women	27 (41.5%)	1 (1.5%)	1 (1.5%)	3 (4.6%)	4 (6.2%)	24 (36.9%)	5 (7.7%)	65
Men	16 (53.3%)	0	0	3 (10%)	1 (3.3%)	6 (20%)	4 (13.3%)	30
Total	43 (45.3%)	1 (1.1%)	1 (1.1%)	6 (6.3%)	5 (5.3%)	30 (31.6%)	9 (9.5%)	95

$\chi^2 = 5.358$ (p-value = 0.499).

Table 5.13 Older versus younger people's motivations

	F	FG	Fr	G	O	S	U	Total
Old	33 (30.6%)	1 (0.9%)	4 (3.7%)	11 (10.2%)	4 (3.7%)	44 (40.7%)	11 (10.2%)	108
Young	43 (45.3%)	1 (1.1%)	1 (1.1%)	6 (6.3%)	5 (5.3%)	30 (31.6%)	9 (9.5%)	95
Total	76 (37.4%)	2 (1.0%)	5 (2.5%)	17 (8.4%)	9 (4.4%)	74 (36.5%)	20 (9.9%)	203

$\chi^2 = 6.741$ (p-value = 0.345).

Table 5.14 Are older believers voluntary workers?

	Buddhism	**Christianity**	**Falungong**	**Taoism**	**Total**
No	25 (69.4%)	17 (73.9%)	1 (100%)	44 (91.7%)	87 (80.6%)
Yes	11 (30.6%)	6 (26.1%)	0	4 (8.3%)	21 (19.4%)
Total	36	23	1	48	108

$\chi^2 = 7.51$ (p-value = 0.057).

Table 5.15 Are younger believers voluntary workers?

	Buddhism	**Christianity**	**Taoism**	**Total**
No	19 (90.5%)	30 (68.2%)	29 (93.6%)	78 (81.2%)
Yes	2 (9.5%)	14 (31.8%)	2 (6.5%)	18 (18.8%)
Total	21	44	31	96

$\chi^2 = 9.184$ (p-value = 0.010).

The number of men and women in both groups who volunteer is not significantly different: 20 per cent of older people do voluntary work and 80 per cent do not, while 18.8 per cent of younger people do and 81.2 per cent do not. Table 5.14 shows that most older Taoists are not voluntary workers (only 8.3 per cent), while about one-third of Buddhists and Christians are. Table 5.15 shows a different distribution: younger Christians are more likely to do voluntary work than are younger Buddhists and Taoists. This suggests that Buddhism and Taoism may need to change their strategies to recruit more young voluntary workers.

Discussion and conclusion

Catholicism in particular has been adjusting to Taiwanese culture to allow its believers to continue the tradition of respect for their ancestors. However, Christianity as a whole has not done this. According to government records, by 1995 only 2.7 per cent of Taiwanese religious believers were Catholics. This would seem to indicate that allowing believers the practice of respecting their ancestors has not significantly increased the number of practising Christians in Taiwan. However, the data in Tables 5.1–5.3 reveal a higher percentage of Christians in Taiwan than do the government records. A possible reason for this discrepancy is the research method used here. The educational level of the interviewees for this study would likely be higher than that of the general population since the university students conducting the interviews would be more likely to interact with people of their own educational background and social status.

The fact that more women than men go to church/temple might be explained in one of three ways: (1) Taiwanese women are oppressed; as such, they need more interaction with God; (2) Taiwanese women are less busy than men and have more time to go to church/temple; or (3) Taiwanese women are more sensitive to their emotional or spiritual lives. They have more worries than men, and thus need more of the comfort which religion provides.

Tannen (1986, 1990, 1994) proposes that women like to be connected with people, while men prefer to be independent. Attending church and temple is one way to connect with people or other beings such as God or Buddha. Also, some men might consider it a weakness to admit they need help in dealing with their mental, spiritual and material problems; attending a place of worship may be considered by them to be an admission of such a weakness.

Many Taiwanese reject the passive expression of traditional conservative Buddhism. Young women who want to become Buddhist nuns may be criticized for avoiding the secular world. Men who want to become Buddhist monks are not as likely to be criticized. However, for the past decade, many Buddhists are becoming more actively involved in the world by providing shelter and medical help for victims of disasters.

Tables 5.7 and 5.8 would indicate that Mandarin will become more and more essential in religious places of worship. In Taiwanese mosques today, the Qu'ran is read in classical Arabic, but explained in Mandarin. Many Taiwanese Christian churches have priests, fathers or sisters coming from English-speaking countries. They know English is an international language, and offer English worship. The number of Christians in Taiwan is expected to increase.

The Ministry of Interior Affairs calculated that in 2002, 12.5 per cent of Taiwan's newborn children were born of a foreign parent – mainly from Cambodia, Indonesia, Thailand or Vietnam (Hsue, 2003). The Indonesian Islamic women married to Taiwanese husbands like to gather in mosques to socialize. It will be worth noting if the number of female Muslims in Taiwan will increase and if more Taiwanese people will speak Arabic in the future. It would also be interesting to do a follow-up study 20 years from now, comparing the religious distribution of the young people interviewed for this study to what they are at that time.

Notes

1 The term 'Christian' will here refer to the Catholic, Protestant and Mormon expressions of faith, even though the differences between the three groups are often times deep and complex, with much rivalry within many Taiwanese communities.

2 Sometimes the term 'Taiwan people' better identifies those understood as the Taiwanese (Southern Min speakers who immigrated to Taiwan in the seventeenth century, who claim to speak Taiwanese), whereas other terms refer to specific groups, such as Mainlanders (people from all over Mainland China, who took refuge in Taiwan around 1949 and who regulate Mandarin as the national language), Hakka (the Hakka speakers who also arrived in the seventeenth century) and Aboriginal Tribes (who lived in Taiwan before the arrival of the other groups). For the purposes of simplicity, in this chapter 'Taiwanese' will refer to the general population of Taiwan.

3 In Taiwan, a Taoist and Buddhist place of worship has more than one idol: the main god and others.

References

Campbell, W. (1889). *An account of missionary success in the island of Formosa*. London: Trubner; Reprinted 1972 by Taipei: Cheng Wen Publishing Company.

Chen, M. (1996). Taiwan Han-people believing in Presbyterian at the late Ching Dynasty. In: C. Lin (Ed.), *Christianity and Taiwan*. Taipei: Christian Cosmic Light Media Center (pp. 55–87).

Chen, M. (2000). *2000–2001 Show of western religions in Taiwan*. Tainan City: Tainan Cultural Property Association.

Chuang, Y. (2003). *Ceremonies of sending the evil ships away* (DVD). Taiwan: Center of Traditional Arts and Culture.

Executive Yuan (2005). Statistics. http://tao.mtjh.tp.edu.tw

Hsu, H. (2000). *Records of Lugang town: Religion*. Lugang: Town Hall.

Hsue, C. (2003). Changing Taiwanese families – foreign female parent phenomenon. http://www.npf.org.tw/PUBLICATION/SS?092/SS-B-092–019.htm (as of 10 April 2004).

Huang, S. (1993). *Language, society, and ethnicity: A study on Taiwan sociolinguistics*. Taipei: Crane.

Labov, W. (1972). On the mechanism of linguistic change. In: J. J. Gumperz and Dell Hymes (Eds), *Directions in sociolinguistics*. Oxford: Basil Blackwell (pp. 512–38).

Landsborough, M. (1972). *In beautiful Formosa*. Taipei: Cheng Wen Publishing Company.

Liao, C. (2005). *Jokes, humor and good teachers*. Taipei: Crane.

Liao, C., and Lii-shih, Y. (1993). University undergraduates' attitudes on code-mixing and sex stereotypes. *Pragmatics* 3(4), 425–49.

Shih, Z. (1992). *Words for meditation, Vol. 2*. Tzu Chi wen-hua chu-ban she ('Tzu Chi Publishing Co.').

Tannen, D. (1986). *That's not what I meant*. New York: Morrow.

Tannen, D. (1990). *You just don't understand: Women and men in conversation*. New York: Morrow.

Tannen, D. (1994). *Talking from 9 to 5*. New York: Avon Books.

Wang, C. (1996). Review of Christian universities in Taiwan. In: C. Lin (Ed.), *Christianity and Taiwan*. Taipei: Christian Cosmic Light Media Center (pp. 205–28).

Yisikakafute, Y. (1997). View Ping-pu tribes from aborigines. *Chiao-shou-lun-tan Monograph* 4, 1–15.

6
Women's Letters to the Editor: Talking Religion in a Saudi Arabian English Newspaper

Hannes Kniffka

The letters to the editor (LTE) investigated here appear on a page titled 'Islam in Perspective' in a column titled 'Our Dialogue' in which the Religious Editor of the Saudi Arabian newspaper *Arab News* (and in a similar column in the newspaper *Saudi Gazette*) answers questions sent to him by readers of the newspaper. This chapter looks at a corpus of women's LTE (henceforth WLTE) only. They represent a small fraction (less than about 10 per cent) of a corpus of LTE (of a total of some 150 LTE written mostly by men, analysed in different contexts and respects in Kniffka, 1994, 2001, 2002). All WLTE investigated represent one and the same text type, or 'sub-type'. They all have the structure *Question (Q)*, sent in by a woman author, + *Answer (A)*, supplied by the Religious Editor of the newspaper.

This chapter will focus on which components and interrelations of components in terms of the 'Ethnography of Communication' are to be analysed for an adequate description of LTE as communicative events as well as differences between LTE (or questions in LTE) written by men and LTE written by women.

Some linguistic data, for example text coherence and text cohesion phenomena, the enumeration and 'bundling' of questions by the editor, the ways in which name references to women authors are given or not given in LTE are descriptive data worth being stated in their own right (see Kniffka, 2005). In the focus of this chapter is the question, to what extent and in what way do the linguistic features of the texts reflect the 'socio-cultural position of women', their social status, public role(s) in the Saudi Arabian society of the 1980s and 1990s? Special emphasis is laid on questions that women ask (or are reported to ask according to

LTE appearing in the press) in religious matters, which concern the totality of (their) social and cultural life in this strict Islamic (Wahhabite) society.

Theoretical frame of reference: the 'Ethnography of Communication' components of LTE-communication

From the few questions mentioned above, it follows that a description of the verbal (grammatical) data alone available in LTE published in Saudi Arabian newspapers would be inadequate. This means that a *sociolinguistic* or *sociology of language* framework is needed to account for the complexity of the phenomena envisaged, and to yield an adequate description and explanation of LTE-communication. In other words: the communicative event constituted by LTE on religious matters in Saudi Arabian daily (English) newspapers requires a perspective reaching far beyond grammatical data. This is confirmed by the fact that many salient 'cultural' data are to be accounted for.

The most adequate framework that takes this into account seems to be an adapted version of the 'Ethnography of Communication' model developed by Dell Hymes, John Gumperz and others.[1]

It is not possible here to describe a well-defined set of theoretical postulates, let alone a coherent theory of this type of mass-media communication. A few salient facts on some components and their interaction will be outlined below. Which components are of particular salience for the description and explanation depends on the perspective chosen.

The main task, in essence, is to describe each component of the interaction systematically in its own right. This is not meant as a checklist of single items, but as a description of the 'dynamic' interaction and the complex interrelation of components, so that eventually the total 'dynamics' of the components and the communicative act in its holistic 'Ganzheit' will be accounted for. Due to the questions in focus in this paper, the components setting, sender and addressor, receiver and addressee/audience, message form, message content, genre will be of prime concern. Other components, which are of equal relevance for an exhaustive description, cannot be treated here but just be mentioned in passing, like the 'phenotype' (layout and the 'ensemble' of the page on which the religious LTE appear, Kniffka, 1980), the norms of the interaction and so on. From the basic observable data it can be seen that it is a very complex interaction indeed that is to be described: (published) LTE in the Friday edition of a Saudi Arabian English-speaking newspaper is 'person-to-person' written communication in the first dyad (i.e., between the LTE author and the Religious Editor), and at the same time it is to be described as 'mass communication', since the questions and answers published in the newspaper edition are directed towards a large non-individual readership consisting of hundreds of thousands of readers. The Religious Editor is a complex entity also: he/his team is responsible for the selection of (parts of) LTE which appear in the newspaper edition (and those which do not appear). He is the sender-authority responsible for the actual wording (message form) and the content of a particular LTE and he may have (and very likely does have) a large range of different motivations for making an LTE in the printed version look the way it does; satisfying the expectation of the government, the newspaper owner/editor, the readership and so on. All this is in itself reason enough to choose an 'ethnographic' frame of reference for an adequate description.

In addition, LTE in an English-speaking newspaper in the setting of Saudi Arabian culture is, by definition, 'intercultural communication'. It involves, as a rule, interlocutors from (at least) two different cultures and languages. In fact, very few of the LTE authors (newspaper readers) seem to be natives of Saudi Arabia. The majority of LTE authors (as inferred, among others, from their names and the topics chosen) do not seem to be natives of Saudi Arabia nor native speakers of Arabic. The main readership of the religious page of the English-speaking Saudi daily papers, at least in the time period from which the corpus sample is drawn (1980s and 1990s), seem to represent non-Arabs and non-native speakers of Arabic, but speakers, for example, of languages like Urdu, Pashto, Gujarati, Tamil, Tagalog, Bahasa Indonesia and others.

This can be explained by several factors mainly of the local setting: there is also an Arabic edition of the English daily paper *Arab News*. If native Saudis (and speakers of Arabic) were to write a letter to the (religious) editor of a newspaper, they would very likely choose Arabic (rather than English) in an LTE and would send it to an Arabic-rather than an English-speaking newspaper. Why write a letter to the editor in one's native country in a language other than one's native language?

The socio-cultural setting of Saudi Arabia in the 1980s and the 1990s

A few ingredients of the general Saudi Arabian socio-cultural setting in the 1980s and the 1990s must be mentioned here. The Kingdom of Saudi Arabia is a theocracy of strict Wahhabite denomination. Islam is the state religion, which means that every citizen must be a Muslim (and non-Muslims cannot be given citizenship). The king has, among others, the title of 'custodian of the two holy Harams' (*ÌÁdim al-Îaramayn aš-šarÐfayn*), that is, the two cities (and holy districts) of Mekkah and Medinah forbidden to non-Muslims. Worship of religions other than Islam is strictly forbidden. Christians, for example, who live in the Kingdom as 'guest workers', including engineers, medical doctors, university professors and so on, are not allowed for instance to import a Bible in their personal luggage or to wear a little crucifix around their neck. Non-Muslim foreigners are expected and required to observe general Islamic laws and public regulations. They must not drink, eat or smoke in public during daylight in Ramadhan, must not kiss their wives goodbye at the airport, must not drink or import alcohol and so on. Some regulations seemed less strict in the first half of the 1980s than in the second and had become much more strictly observed in the 1990s due to the increased influence and public appearance of the 'religious police' (*al-muÔawwiÝÙn*). While it was possible (and not explicitly forbidden) for non-Muslim men to swim in short swimming trunks in the Red Sea a few miles outside the city (of Jeddah) and to go shopping in shorts (bermudas or similar) in the early 1980s, this has become more and more frowned upon and is frequently explicitly forbidden now. Needless to say that there are no cinemas or liquour stores around. Saudi men may marry (foreign) non-Muslim women, who, within a short time, will convert to Islam. No Saudi woman is allowed to get married to a (foreign) non-Muslim man, however. Severe penalties exist for violations of Islamic laws as constituted by the Shariah code of justice which has been effective in the Kingdom of Saudi Arabia for as long as the country has existed.

It should be added, however, that within the strict rules of Islamic religion and culture, life was (in the 1980s) enjoyable even for non-Muslim Europeans. Saudi Arabia then was probably one of the safest countries in the world. In the early 1980s, shopkeepers used to leave their stores for prayer (*ÒalÁt*), leaving non-Muslim foreigners in the store together with piles of banknotes unattended on the shop counter. In later years, everybody was asked to leave the store before prayer began. Foreign people, Muslim and non-Muslim, were treated by the unparalleled laws of hospitality everywhere in the countryside. In the *ÎadÐ×* it is stated that a stranger coming to anybody's house is to be treated by the host to the best he can. Only after three days is the host allowed to ask what the visitor has come for.

For the communicative setting in which the newspaper interaction investigated takes place, a few data pertaining to the interrelation of the components *setting; sender; message form; genre;* and *instrumentalities* (channel and media characteristics proper) must be sufficient here (for the components sender and receiver see LTE 8 and LTE 9).

In 1992, there was (according to official statistics by the Ministry of Information) a total of ten Saudi daily newspapers, three of them English speaking, with a total circulation of some 600,000; 34 per 1000 population. Radio (1994): there were some 3,800,000 receivers (1 per 4.6 persons) in Saudi Arabia. Television (1994): some 4,700,000 receivers (1 per 3.7 persons). Radio and TV broadcasting are operated by the Ministry of Information.

Education (Literacy): in 1995, 62.8 per cent of the population aged 15 and over were literate; 71.5 per cent males; 50.2 per cent females. In 1995, Saudi Arabia had a total population of 17,880,000; 82 per cent Saudi, 9.6 per cent Yemeni, 3.4 per cent other Arab, 5 per cent other. Religious Affiliation: Muslim (mostly Sunni): 98.8 per cent; Christian: 0.8 per cent; other: 0.4 per cent (1983).

In 1965, King Feisal ordered television to be installed, after a painstaking discussion process with opponents, in particular the Ulema (the members of the highest religious council). The programme consisted of Qur'an recitations and other religious broadcasts, yet the opposition to the media was not silenced – leading to the tragic events that finally led to King Feisal's assassination in 1975. Today, there is still ample coverage of religious topics in both official Saudi TV channels, channel 1 (broadcasting in Arabic) and channel 2 (broadcasting in English). In the early morning and at bedtime, Qur'an verses are recited. During *ÒalÁt*, five times a day, TV programmes are interrupted and a (steady) picture of the *KaÝ́ba* is shown. There is also entertainment (mainly Egyptian and Indian films, US soap operas). Another important local fact: public cinemas and so on have been forbidden in Saudi Arabia since the late 1970s. At the same time, the Saudi population seems to be one of the best equipped in the entire world with videos of various denominations to be shown in private, which again is difficult to analyse in reliable scientific terms for obvious reasons. TV programmes, sales of videos, any public communication whatsoever is governed by strict (official) censorship.

Two of the three English-speaking daily newspapers, *Arab News* and *Saudi Gazette* appear in Jeddah (the third, the *Riyadh Daily*, published in the capital since the late 1980s will be left out of consideration here). *Arab News* has a circulation of some 53,000, *Saudi Gazette* about 35,000 (1996). For comparison, some data on German daily papers: in Germany, 375 daily papers with 1600 local editions and a total circulation of some 26 million copies are published daily (in addition

28 weekly papers with 2.2 million copies and 8 Sunday papers with 4.8 million copies), that is, 314 copies per 1000 inhabitants. Four-fifths of Germans aged 14 and over (80.7 per cent) read a daily paper (50 million). Some 16.4 per cent read 'their' daily paper. Reading habits: 78 per cent read the local news daily, followed by political news, foreign policy. Of particular importance to the readers are ads, editorials and letters to the editor. The last mentioned fact of reading behaviour seems to be quite similar to readers of the two English-speaking newspapers in Saudi Arabia.

The *Arab News* and *Saudi Gazette* have been running personal ads since about the mid-1980s on topics like cars for sale, shared housing and rooms for rent which, being novelties then, people were rather eager to read. The *Arab News* has been running LTE at least since the early 1980s on all kinds of secular (culturally permitted) topics and in its Friday edition runs a page 'Islam in Perspective', a 'religious page' so-to-speak, with a column 'Our Dialogue' in which both editorials by Islamic scholars and a large section of LTE appear, the latter in the form of questions sent to the Religious Editor of the paper and his answers to the readers. *Saudi Gazette* has followed this example since the 1990s (they carried no such coverage a decade earlier).

There is one other fact of the general cultural and religious setting worth mentioning here: the different degree of tolerance Saudis (strict Muslims) display against the two other monotheistic revelatory religions, Christianity and Judaism, on the one side, and polytheistic and other 'religions' on the other (Saudis would not use and accept this term for them), like Hinduism, Buddhism, Jainism and others.

There is also racial and/or ethnic prejudice in Saudi Arabia, though not as overtly marked and perceivable as the former. Saudi hospitality, students' genuine friendliness towards their teachers, be they locals or from abroad (as put down in the Qur'an), an enormous spontaneous personal appreciation of others, all this is activated almost self-evidently towards other Arab Muslims and, for example, German *'ÌawÁÊa'* – not so towards people from India, the Philippines, China and so on, even if they are Muslims. A Saudi MD, asked about his attitude towards black people in general and (Saudi) fellow citizens in particular, would echo the general official version that they were, of course, fully recognized, that there was no discrimination, no antipathy and so on. When asked in private, he would say that he would never allow one of his daughters to get married to a black Saudi. There is an (entirely un-Islamic) hierarchy of appreciation for other races and ethnicities with white/Caucasian people from the United States and Europe on one end of the scale and people from India and the Philippines on the other.

Corpus of women's LTE

The textual data of the questions of the WLTE investigated are given in Table 6.1.

In the left-hand column, the name and the edition of the paper is given, in this case, different editions of the Saudi Arabian newspaper *Arab News* (AN) of 1987 and of 1996 and one edition of the Saudi Arabian newspaper *Saudi Gazette* (SG) of 1997. The right-hand column contains the original wording of the LTE as it appeared in the newspaper. The 'headlines' (HL) of the LTE and also the numbering and compartmentalization of questions (or the lack of it) were supplied by the Religious Editor. There are no explanations for why the editor chose this particular text-compartmentalization or why these headlines were chosen.

The selection of the corpus of WLTE in Table 6.1 as a whole is accidental. The overall criteria for selecting these WLTE and not others is that the 'name line' underneath the LTE as it appears in the newspaper and/or the text is an explicit indication that the LTE author is a woman rather than a man. There is in fact a continua of explicitness (or non-explicitness) in this respect (Kniffka, 2005), meaning that in some cases the form of the name given does not identify the sex of the author, but, as in LTE (3), the wording and the content of the LTE do, 'my husband seems

Table 6.1 Women's LTE

LTE (1)	
AN, 08.11.1996	**Inheritance that was usurped** Q1: When our father died 20 years ago, he left behind his wife, 4 sons and 5 daughters and a family house. One of our brothers has died before our father. Our two eldest brothers sold the family house 10 years ago and used the money in their business. Our mother died 2 years ago. Now our two brothers have given each of us his or her share of the original price they received for the house. They also paid the share of our deceased brother to his widow, not to his four children. Our father did not make a will in favor of those because they are well-off. Please comment. Q2: What is correct English spelling of the name of the mother of prophet Ishmael. Is it Hajra or Hajer? Does it have any meaning? Q3: Can a Muslim study American law and practice law in U.S., accepting only the cases that do not involve anything against Islam? Mrs B. Sheikh, Jeddah

Continued

Table 6.1 Continued

LTE (2)	
AN, 01.11.1996	**How should we react when others do wrong?** Q: After the death of my husband I took my children to settle down at another place. When I came back, I found someone occupying our house on the pretext that my late husband owed him money. I agreed to pay in due course, but he wants immediate payment. Can he occupy the house in this way? Similarly a motorcycle which belonged to my husband was sold on installments but the buyer paid only the first installment. Is it lawful to delay payment in this way? Affected Lady, Makkah
LTE (3)	
AN, 09.08.1996	**A second marriage** Q: After 10 years of marriage, during which I have had four children, my husband seems intent to marry a second wife. My protestations have gone to no avail. I am a working woman and my husband takes away all my salary. In his attempts to get married, he is telling other people that he is unmarried. I am very worried about my life and children. Please comment. E.A.R., Riyadh
LTE (4)	
AN, 18.10.1996	**What to avoid in the family home, etc.** Q1: If a wife is disobedient to her husband, is her prayer acceptable? What if a man is married to a domineering wife and he is scared of contradicting her? May I also ask whether it is true that most women go to hell? Q2: Does Islam allow women to work in factories or offices? Q3: Can a man marry a second wife without the approval of the first wife? H. Nazeer (Mrs), Jeddah
LTE (5)	
AN, 22.11.1996	**Briefly** Q1: If at any place the period between dawn and sunset lasts for 18 hours or so, what is the time when Muslims should finish their fasting? Q2: Is it permissible to eat kosher meat? Q3: Why are stones exempt from zakah when they have considerable value? N. Jaffri (Mrs), Michigan, USA

Continued

Table 6.1 Continued

LTE (6)	
AN, 20.09.1996	**Woman's appearance and dress** Q1: Is it permissible for a young woman to shape her eyebrows, particularly when they are too thick or irregular? What about the removal of facial hair, particularly over her upper lip? Its presence is likely to invite ridicule. Q2: Is it permissible to replace hair if one becomes bald at a young age? Q3: Is there anything wrong with a girl wearing loose pants which are more comfortable particularly during travel? M.M.A., Riyadh
LTE (7)	
AN, 04.05.1987	**Can one take medicine while fasting?** Q: As a doctor, I would like to know whether intramuscular and intravenous injections may be given to a fasting person, in normal circumstances or in emergency. Can a person who suffers from asthma use his aerosol inhaler while fasting? What is the ruling regarding the use of ear, nasal and eye drops, suppository and the drawing of blood for investigation during fasting? Naseema Ismail, Riyadh

intent to marry'. In some cases there is neither any explicit 'text-external marker' in the name line nor a 'text-internal' marker identifying the sex of the author, but just some degree of likelihood that a woman may be the author. In LTE (6) the content of the three questions asked about 'women's appearance and dress' seems to be more located in the domain of women than of men.

The total amount of LTE-data of the period between about 1981 and 1999 available to me suggests that there is a high percentage of recurrence of wording and structure, and also of topics of LTE. One can say that the topics of the LTE in 1983 are more or less the same as the topics in LTE in *Arab News* in 1993 and 1996. This in itself is an interesting culture-specific fact of Saudi Arabia. It does not indicate that there has not been a development concerning the role of women in Saudi Arabia. It rather suggests that there has been a certain continuity in the way LTE authored by women on religious topics have been handled by the Religious Editors of the newspapers during these two decades. Some minor changes certainly did take place in details in the 1990s, such as the fact that the Religious Editor of *Arab News* is mentioned by name at the head of the page written with his by-line right below the page's title 'Islam in Perspective', which was not customary during 1981–8.

The topics occurring in the LTE investigated can be grouped together into a limited amount of classes, which also reflects a relatively high degree of recurrence of topics on which women write LTE and the Religious Editor chose to answer in the Friday edition of the newspaper. The most frequently asked questions involve:

- Inheritance
- Second marriage of husband
- Women's 'proper' appearance and dress
- Execution of religious life and duties
- Equal rights and/or value of women
- The role of women in congregational prayer
- The general role of women in Islam
- Proper spelling and proper language use (of Arabic, English and other languages)
- Legal aspects of Islamic life abroad in a non-Muslim country
- Selected special questions asked by women, but referring to general, other than women's concerns.

Sociolinguistic analysis of LTE written by women

The components 'message form' and 'genre' of LTE

It is worth noting that the reader of the newspaper, is not in any way in a position to judge which 'message form' the portion of the text printed in the newspaper is created by the LTE author and which is due to modifications by the Religious Editor. The only persons to know would be the author and the editor. Such details are generally not revealed to people other than members of the newspaper's staff, in Saudi Arabian as well as in Western newspapers. As a rule, newspapers keep this information to themselves, just printing a certain selection of LTE in a paper's edition, without saying which LTE or parts of LTE were not printed. The actual wording of the LTE as a printed text product appearing in a newspaper is a complex interrelation of components and sub-components, here mainly of the sender, the addressor, the message content, the genre, the receiver, the addressee, the audience and the message form.

In *Arab News* the page 'Islam in Perspective' has a little box addressed 'To our readers', stating the postal address of the paper and its Religious Editor, to which readers should send their LTE.

Another ingredient worth stating for message form is the ensemble of the whole page 'Islam in Perspective' containing several texts on Islamic topics, stories, a calendar of important historical Islamic events, the column 'Our Dialogue', and also the fact that newspapers like *Arab News* and *Saudi Gazette* have secular (non-religious) LTE on different topics. They appear (in the 1990s) on Wednesdays in *Arab News* and are differentiated from the religious LTE other than by the day of appearance alone, but also by the place/page where they appear in the newspaper edition (on a page other than the religious page), text type, topic (that is non-religious topics, such as politics, sports, domestic affairs) and phenotype/make-up. A more detailed analysis of the components cannot be given here. The general descriptive and explanatory potential seems clear.

The components 'sender', 'addressor'; 'receiver', 'addressee' and 'audience' in WLTE

As mentioned above, the text type 'letter to the editor' is primarily characterized by the specific complexity of the component's 'message form'; and of 'sender', 'addressor'; 'receiver', 'addressee' and 'audience' and their interrelations.

The 'sender' component itself is characterized, *inter alia*, by the fact that it is a 'global' component, consisting of the following constituents:

- The newspaper editor(s) and the other members of the newspaper staff as a whole.
- A particular, identifiable sender, the Religious Editor. He also is an 'addressor', that is, the text author of the 'answer-portion' of an LTE published in the paper. To what extent he is also 'addressor' ('editor') of the 'question-portion' of the LTE cannot be determined accurately.
- A single, identifiable 'addressor', the author of the LTE (here of the questions asked in the LTE).

This structural complexity of the sender component has no direct equivalent in the 'receiver' and 'addressee' categories. The latter are characterized by the general fact that there is no personally identifiable receiver/reader of the newspaper, but a global addressee and audience category, unless s/he acts, in turn, as a sender and writes an LTE, which gives it a unique structure of its own. LTE are the only media or the only text type (apart from ads) in which readers of the newspaper have the possibility to act as senders (authors) in the newspaper. So there is the rare chance that text products of the readership which would not be heard or read in print otherwise can be accounted for. One can say that in mass-media communication Hymes's distinction between 'addressee' and 'audience' within the receiver category coincides or is 'neutralized', with the exception of LTE, which makes this another definitory trait of the text type LTE.

The category of 'addressor' as it appears in the LTE printed in a Friday edition of a Saudi Arabian newspaper carries highly significant socio-cultural information about the interaction as such, about society and about the people involved.

All 9 LTE of the corpus were written and sent to the Religious Editor by *women* authors, which is more or less explicitly stated in the 'name line' at the bottom of the LTE (see Table 6.1).[2] The most general opposition is between male authors of LTE on the one hand, whose names are given by FN + LN + CN.[3] FN are sometimes abbreviated/given in initials only, with the general understanding, that the author of the LTE is a male (the 'unmarked case' of sender- and authorship indication), and LTE authored by women on the other hand, which have an addition *(Mrs)* (mostly in parentheses) added to the name (as the 'marked case'). In respect to sender-identification, it is noteworthy, too, that no LTE with an addition **(Miss)* or **(Ms)* respectively occur in the corpus. This could mean that LTE sent in by female authors are written by married women only, that the Religious Editor, or the editor of the paper chooses the variant *(Mrs)* only, no matter if the women authors are married or not, or that it is the paper's policy to print LTE authored by married women only. There is no way of telling which is the correct explanation here. The fact that women authors are identified by the addition of *(Mrs)* is no doubt a basic marker of the structure of Saudi Arabian society and the position of women in it.

There is an interesting difference between a particular LTE versus the rest of the LTE of the corpus: In Table 6.1 LTE (2) it is not FN + LN + (Mrs) + CN that is given, but rather an anonymous, more general reference by an appellative *Affected Lady, Makkah* (I have not seen an example such as **Affected man, Jeddah*).

Presumably in all newspapers, the names of the authors of some LTE are withheld; that is, they do not appear in print, though the editor may know (or, frequently, is required to know) them. Some papers, including the two Saudi Arabian papers investigated, have the policy to accept LTE only with the full name and address of the sender given to the editor. They may or may not appear in print. Seemingly one of the major reasons for this is the specific status, maybe 'touchiness' of the topic (manifest content) which the LTE in question has. In the case, for example, of LTE (2), the author may not wish to be identified by name because the LTE refers to a 'touchy' matter and/or the LTE-editor may not wish to identify her publicly. As with other 'internal' editorial data, the two alternatives cannot be determined any further on safe empirical grounds by the linguistic observer. That an editor and/or Religious Editor does not want the name to appear in print may itself be due to a large variety of reasons. The author may not want to identify herself for different reasons, too. There may also be 'joint ventures' of one kind or another. There is no way of telling which (combination of the two) applies. Interestingly enough, one may thus be able to state a special 'status' of the topics in such LTE in which the name of the author is not revealed without being able to come up with any detailed valid explanation. The same applies basically to LTE written by men (cf. Kniffka, 2002). In a particular LTE a man may inquire, for example, about unislamic behaviour or some wrong-doing, and not surprisingly does not want to identify himself.

The 'touchiness' of a particular topic is itself an interesting indicator of culture-specific constraints: What is 'touchy' and what is not is highly marked by culture-specific factors; what one is eager or ready to talk about in (the) public (press), even more so. There is no overall intercultural invariant hierarchy of which topics are 'touchy' in a culture x and which are not. On the contrary, its status is by definition a highly culture-specific matter. Some topics may be touchy, dangerous or even forbidden, or severely punishable if expressed in public (e.g., mass media) in Saudi Arabia, but are no problem at all if publicly stated in the West.

In the case of LTE (2) in Table 6.1, the topic as such does not seem in any conceivable way to be a topic which would suggest or require withholding of the name from a Western European perspective. The 'Affected Lady' in this LTE complains that, when she had moved to another place, someone occupied her house because her late husband owed the usurper some money. So she asks the LTE-editor whether the man has any right to occupy her house and also whether it is lawful that he has not paid the total instalments for a motorcycle that her late husband had owned (his name is not mentioned). It should be stated here, that in none of the other questions of the corpus are the name and address of the sender/ author withheld. Neither the LTE with a question like *Is it true that most women go to hell?* nor *Can a man marry a second wife without the approval of the first wife?* nor *Is it permissible to eat kosher meat?* are anonymous. Obviously Saudi culture views some matters differently from Western societies. Some questions would not likely be asked in a Western daily paper (for a variety of reasons); some would neither require nor ask for the withholding of the author's name (as here in LTE (2), 'collecting debts').

Culture-specific facts of the setting of women in Saudi Arabia and in WLTE

Some of the more important data of the general socio-cultural setting of Saudi Arabia have been outlined above. Below, the data will be supplemented by those pertaining to women in Saudi Arabian culture in the 1980s and 1990s. Special reference will be made to the LTE written *on* women and *by* women. As was also stated, Arabic-speaking Saudi newspapers address a readership different from that of English-speaking Saudi daily newspapers (and certainly weekly papers). Not many native speakers of Arabic (and Saudis for that matter) will read the English edition of *Arab News*. Also, one can tell from many of the names of the LTE published that they are frequently from Islamic, but non-Arabic-speaking countries like Pakistan, Indonesia and others. It is safe to say that the English edition of *Arab News* is definitely a paper for expatriates (Muslim and non-Muslim) living in Saudi Arabia, rather than native Saudis. In addition to the names, the topics of the questions asked and the way they are asked in some cases also reveal that it is unlikely to have a native Saudi author. A native Saudi would mostly not ask such questions, or at least not publicly in a daily paper.

It would also need further detailed empirical study to determine to what extent the range of topics published in a paper reflect the actual interest of the readership and/or the religious convictions and publishing and editing practices of the Religious Editor. It would require an insider to arrive at any real reliable data. Nevertheless, two very basic facts about the LTE are beyond dispute:

1. The selection of WLTE actually did appear in press on the page 'Islam in Perspective' in the column 'Our dialogue' on the dates given. The data are authentic 'real life' data.
2. The WLTE were selected for publication (over others) by the Religious Editor of the paper, that is, they were considered to be of general interest to a large readership, and therefore (and because of additional concerns) were selected.

One can assume that the LTE investigated contain salient information on Islamic life and culture in Saudi Arabia in the 1980s and 1990s. The data are genuine and 'psychologically real' information on what was going on in Islamic matters in Saudi Arabian print media in that period. They are cultural icons of Saudi Arabian Islamic life and culture. They also reflect mental activities, ways of thinking, value systems and attitudes of expatriates living in Saudi Arabia and of at least one Saudi national, the Religious Editor.

A few more facts on women and women's behaviour in everyday Saudi culture must be added. Normally women are excluded from public life in the Saudi socio-cultural setting. They are totally unrecognizable in public and wear wide, long, black clothes covered with an *abayah* down to the ground not revealing any body form. Normally, native Saudi women are veiled entirely in public and can be identified as such by this very fact. Women have their own entrances in banks, hospitals and so on, their own entrance to their house or building. They also have their own living quarters in a flat, including a bathroom reserved for women (and children). This part of a flat/house is not entered by male visitors. Women are not allowed to drive cars in Saudi Arabia, nor to sit next to someone in a taxi, in a car, at a bench, at the beach and so on, who they could possibly be married to.

Taking all this into consideration, it appears 'logical' that, in Saudi culture, women are the 'marked' class of sexes, for example by the addition of *(Mrs)* as senders of LTE, whereas the 'unmarked' forms (e.g., initials for FN and a LN) refer to men as LTE authors.

A brief comparison of men's LTE (cf. Kniffka, 1994, 2002) with the WLTE of the corpus shows that nearly all topics and topic areas occur in both (LTE by women and by men), notably with the exception of 'Equal rights and value of women' occurring *only* in WLTE, and some obvious specifications, like 'growing moustache and beard' in men's or 'women's proper dress' in women's LTE. A brief look at topics of men's LTE (taken from Kniffka, 2001) continues this:

- Problems of marriage, second marriage, divorce, sexual and domestic behaviour.
- The role and status of women at home and in public.
- Questions of Islamic law in terms of civil rights, *zakah* (alms), death, inheritance.
- Performing the daily religious duties of a Muslim properly contained in the so-called 'five pillars of Islam' (five prayers per day, fasting, *zakah* and so on).
- Questions of proper Islamic behaviour, including linguistic questions like proper spelling of names, pronunciation, reciting the Qur'an and so on.
- Proper behaviour when living in a non-Muslim environment, like, for example, professional conduct as Muslims in the West, working as a lawyer in the US; which direction to bow to when offering prayers if, for example, living in the south-east of Canada or in South America or aboard a plane.

In other words: with the exception of a (significant) difference in one topic, a large symmetry or overlap can be shown for men's and women's LTE. This means that, with the exception mentioned, there are no or few LTE-topics 'specific to women', nor LTE-topics 'specific to men', but rather a common class for both. This does *not* mean, that LTE written *by* men and LTE written *by* women are alike, however. There are differences in the frequency of occurrence of some topics in each, such as problems of 'second marriage', which are, not surprisingly, more frequent a topic in LTE by women, zakah and other financial matters being more a topic in LTE by men.

The large coincidence of topics and the somewhat 'limited' scope of questions asked as such is itself a culture-specific fact of Saudi culture, and seems, above all, a characteristic of the 'ensemble' given by the page 'Islam in Perspective'. At the same time it reflects the usages and editorial policies of the Religious Editor.

It is safe to say, that all this is a sign of a relatively strong impact of Islam on Saudi Arabian culture, affecting everybody, men and women, in basically the same fashion and strength, modelling the differences of both as defined in Islam in a clear cut way. This is in fact topicalized quite frequently in (W)LTE.

Two exemplaric LTE written on women and by women

It was pointed out briefly, that the LTE written *on* women (by men) and the LTE written on women *by* women do not seem to contrast significantly in terms of topic and wording. On the contrary it is plausible to assume that both are very similar, as far as the **questions** of the LTE are concerned.

A different matter may be the answers supplied by the Religious Editor. The answers show a considerably sharper contrast for different LTE than the questions and probably also for men versus women as LTE authors. A striking contrast in the answers is found, however, in the LTE investigated in two examples (8) and (9), in which the Religious Editor answers LTE written by women of a different cultural background. The fact that both LTE authors are women is beyond doubt according to the data supplied in the LTE. All other information cannot be stated with equal safety. It is remarkable in the answers of the Religious Editor what differences the texts in both cases contain. It seems worthwhile to give the questions and the answers in full detail.

LTE (8)

AN, 06.12.1996 Equal rights or equal value for women

Q: In a recent reply to a reader's question you have stressed the equality of men and women. May I suggest that this frequently asked question results from a problem of linguistics. The Oxford English Dictionary lists 12 definitions for 'equal'. The first two are 1) being identical in value, and 2) having the same rights and privileges. Both these meanings are common in the everyday usage of education. Thus the two sexes are unequal in the sense of the second definition. You have mentioned on more than one occasion some of the differences between the two sexes. May I suggest then that the answer to whether Islam considers men and women equal should be both 'yes and no'. This will stress their equality in value, but not in rights and privileges.

Dale McIntyre, Dhahran

A: The case is certainly well argued by Mrs McIntyre, but I still disagree with her. People differ in their abilities, aptitudes and temperament. When the law gives certain rights and privileges to all people, they do not exercise these rights in the same way. They cannot, even if they try. Some are bound to have much less than others. It can be argued that the law has given them the chance to be equal, but they cannot make that equality physically and materially apparent. Maybe this is part of what gives human life its richness. But if the law assigns the same rights and privileges to all people without taking care to favour some less endowed or less able groups, these may be at a great disadvantage.

Take the example of education where the law in most countries gives all children the same rights. If the law does not take care to give special facilities to children with special needs, then these children will not have the same education. If the law gives them that, then it appears to indulge in favouritism. Dyslexia is a stark example. It signifies a range of learning disabilities that have no apparent cause and no cure. Yet many children are dyslexic and need to learn special strategies to get round their difficulties and acquire the learning to which they are entitled. Dyslexic children should be given more time on their tests and for their assignments. If the law gives them that, it appears to favour them, while if they are not given extra time, they show themselves less capable, and they are, as a result, at a great disadvantage. To my mind, equality cannot be administered unless these children are given the facilities necessary to learn at their own pace, and also to show their ability. Otherwise, the law which guarantees a minimum standard of learning for everyone cannot be enforced.

Men and women are equal in God's sight, both in value and in their rights and privileges. How can we say that when there are several areas in which they appear to be in a position of inferiority? My reader lists quite a few of these, including that they are not allowed to marry four husbands, and they cannot divorce at will, as well as their inheritance and the fact that in certain cases two female witnesses may serve in place of one male witness.

But if we take these at face value, we will be doing the same as one who claims that God has threatened those who pray. To justify his claim, he quotes the fourth verse of Surah 107, which says: 'Woe to those who pray.' If you take this verse alone without reading the following one which qualifies it, you will think that people should not pray in order to spare themselves the woe with which they are threatened. But if you read the rest of this short surah, your conclusion will be totally different because it reads: 'Woe to those who pray but are heedless of their prayers; who put on a show of piety but refuse to give even the smallest help to others.'

In order not to make such a hasty and faulty judgment, we should carefully consider these differences in the rights and privileges of men and women. When we do, we are bound to conclude that the differences in rights and privileges do not have any bearing on the equality between them. They are meant only to help each of them fulfil the role assigned to them so that both give to human life the best they can. To satisfy ourselves of this basic equality, we need only to remember that both men and women have the same duties to believe in God after reflection and consideration, and to worship Him in the same manner. Both will have the same reward for any act they do in fulfilment of their religious and community duties.

LTE (9)

SG, 03.01.1997 **Can a woman lead congregational prayer, or be Shariah court judge?**

Q: I have heard that a woman cannot lead the congregational prayer or deliver Friday sermons, whereas a man can. Why? Can a woman be a judge in a Shariah court? If not, why? Please refer to some *ahadith* in support of your answer.

Safia Iqbal, Makkah

A: It is a prerequisite for the correctness of the *Imamah* [leading salaah in congregation] men and women that the Imam be a true male. A hermaphrodite may not lead the congregation of males. This cannot be done either in the obligatory (*fard*) nor the *sunnah salaahs*. However, womenfolk may lead only womenfolk in the salaah. There is a hadith from Ayesha as well as from Umm Salmah and Ata that the prophet (*pbuh*) said: 'Truly, the woman leads the *saalah* for womenfolk.' Imam Daraqutni reports from Umm Waraqah that the Prophet (*pbuh*) had someone to make the *Adhan* for her so that she could lead the *saalah* for the womenfolk in her household. This is the ruling of the three Imams. However, Imam Malik does not even allow womenfolk to lead the *saalah* for womenfolk. Over and above this fact, a woman's voice, in as far as reading the Qur'an and singing is concerned, is also regarded as *awrah*, but not her talking. Should she, for instance, lead the *saalah* and have to recite the Qur'an, her voice which is part of her *awrah* would immediately disqualify her and it would also evoke mixed emotions from male worshippers. A man's voice is not regarded as *awrah*. A woman may also not be a judge and a woman is not allowed in some cases to bear testimony on certain matters. One may ask, why? The answer to this is very simple and very logical. As far as being a judge is concerned, he must have all his faculties intact and he should be well versed with what goes on outside, in the marketplaces, the mosques, in fact everywhere, for he should investigate these. A judge needs to have discourse with other jurists on matters on which he is to base his judgment, he needs to speak to the witnesses and everyone participating in the dispute in order to really do justice. It is forbidden for a woman to do all this (i.e., sitting with strange unrelated males due to fear of temptation). Also a woman, despite her high degree of intellect and the high degree of her knowledge, yet she is still governed by a natural instinct, namely her deep compassion bestowed on her by the Creator and she can do nothing about it. And because of this deep compassion of hers, she in most cases may have her judgment clouded. Imagine what the situation would have been if a woman who is a judge is in need to decree the death penalty on someone. In most cases this would be impossible for her to do, because of her natural compassion which will cloud her logic. In the majority of families, this comes to the fore very strongly where children are concerned. Irrespective of their mother's high intellect, they know how to get to her unlike the firm father. Let us take a baby, for example. This child may cry for months during the night, yet the mother despite all the sleep she has lost, would not utter a word of despair or say anything derogatory about the child in frustration. The father would lose his cool after a day or two because again, that is the way Allah created him. He must support the family, he must go and work, he needs his sleep, therefore his reasoning is very logical. Although the mother also needs her sleep, she has been strengthened by Allah with inner strength and with such compassion which overrides every rule of logic. For that reason He said to us in Chapter 4, Verse 32: 'And do not wish for those things where Allah has bestowed His gifts more freely on some of you than on others. To men is allotted what they earn and to women is allotted what they earn. But ask Allah for His bounty, for Allah has full knowledge over all things.' When she has to bear testimony, she is allowed to do so in matters of trade and personal law only. In these cases two males are necessary, but if two males cannot be found, then one male and two females. This is based on the fact that womenfolk generally do not participate in such acts. Then there are situations where womenfolk are even disallowed to be witnesses in court when it comes to accusing two persons of adultery. Here four men are required to testify in front of the judge what they have seen, how did they see it, what positions they were in, were they naked, etc. The entire scenario must be sketched in front of the judge. Allah through His infinite mercy does not allow a woman to witness such filth because she is sacred. However, among all the jurists, it is only Imam Abu Hanifah who allows a woman to judge only on matters where she is allowed to bear testimony and nowhere else.

A brief contrastive interpretation of both texts must do here. As mentioned earlier, LTE (9) is written by *Safia Iqbal* from *Makkah*. The name indicates that the author is a woman. The address (*Makkah*) indicates that she is a Muslim (as stated earlier non-Muslims are not allowed to enter the two Holy Cities of Makkah and Medina). The name *Safia Iqbal* is not a clear indicator of which ethnic background the writer comes from. The name itself occurs frequently with Pakistanis, so the author could be from Pakistan, but could also be from another Arab country or even be a Saudi naturalized citizen. This is not of importance here. More so is the fact that a native Saudi woman would probably not ask the questions that the author of LTE (9) asks. One can assume that a Saudi woman knows and/or would not ask why a woman cannot lead a congregational prayer of men and why she cannot be a Shariah court judge. A different question is whether if she really did not know, she would ask it in public, in particular in an English-speaking newspaper. This seems unlikely. There is no definite proof, however, that this is or is not so.

The name of the author of LTE (8), *Dale McIntyre* from *Dhahran*, indicates that the author is a woman with an English (US or British) name and background, presumably from an Anglo-country and not a native of Saudi Arabia. It is not definitely clear whether she is a non-Muslim or a Muslim, though the former seems much more likely.

In terms of the textual status of the questions in LTE (8) and (9), it is remarkable that LTE (9) asks very specific, rather brief and clearly stated questions, whereas in LTE (8) there are no questions at all, but indirect speech acts instead (though the Religious Editor (henceforth RE) puts a

'Q' for question in front of it). Obviously the LTE author does want to get the RE's comment. The reason for lack of specific questions is again that the text is rather complex.

Another contrast of the two authors of LTE (8) and LTE (9) is that the first is, in simplified terms, 'a Western woman', the latter 'an Eastern woman'. It seems that this is, together with Muslim versus non-Muslim, the prevalent contrastive perspective under which the answers of the RE are to be interpreted. The answers are about equal in size in LTE (8) and (9) (the answer to the Eastern Muslim woman is only slightly longer).

Below, a brief contrastive analysis of the two texts, LTE (8) by a Western (probably non-Muslim) woman and LTE (9) by an Eastern Muslim woman will be given in a very informal way, focusing: (1) on some data of 'message form'; (2) 'message content'; and (3) text pragmatic 'function', the latter pertaining to the argument structure and the way the RE proceeds to get his points across.

As has been mentioned, the questions themselves contrast in form and overall meaning: the question in LTE (9) by an Eastern Muslim woman actually states two precise detailed questions: *Can women lead the congregation of prayer or deliver Friday sermons? If not, why?* and *Can a woman be a Shariah court judge? If not, why?* The fact that *why*-questions are among the most frequently asked in LTE altogether is in itself an interesting characteristic of LTE. LTE (8) by a Western, probably non-Muslim, woman does not have specific questions, but rather offers comments on the RE's and of course Islam's view of men and women. She is not asking direct questions nor asking for a comment explicitly (as in more complex LTE), but is making an indirect speech act *May I suggest then that the answer to whether Islam considers men and women equal should be both 'yes and no'. This will stress their equality in value, but not in rights and privileges*. This is an indirect request for a comment by the RE.

A striking difference as far as vocabulary and more general lexical data are concerned, is that in the answer to LTE (9) towards a Muslim woman, not surprisingly, a lot of Islamic religious terminology (if not jargon) in Arabic occurs, whereas in the answer to LTE (8) directed towards a non-Muslim reader, there is no word borrowed from Arabic in the text. There, the RE uses English medical terms (*dyslexia, dyslexic*) in an argument to support the notion of inequality of people. Obviously the RE wants to suit the addressees and makes full use of his knowledge of terminologies (maybe also to impress his readership).

The abundance of Arabic religious terms in LTE (9) may, in part, have been triggered by the LTE author (*Please refer to some ahadith in support of your answer*).

It is very likely that the Arabic words are understood by parts of the readership (in different depth and to a varying degree): words like *Imam, salaah, hadith* and others may be widely understood, even with non-Arabic speaking non-Muslims. Other terms like *surah, adhan* and *awrah* may not be. This shows that the RE is not really too much concerned with the general comprehensibility of his answers towards the total readership, but more with his LTE author and her question.

This may also be one of the main reasons for the fact that the RE goes into some detail to contrast the various opinions of different Islamic traditions and schools, introducing several names, with which non-Islamic readership may not be familiar.

A rather unusual or even 'modern' way of speaking is revealed by the fact that the RE introduces (as some kind of a special rhetoric device) the term *hermaphrodite*. The controversial opposition of men and women in leading the congregational prayers is somewhat neutralized argumentatively by the introduction of *hermaphrodite*: the RE states that only a *true male* can lead congregational prayer for men, not a hermaphrodite and not a woman. It seems that this rhetorical device is successfully used to minimize the controversial issue of women not being allowed the same functions as a man in leading congregational prayer.

Interestingly enough, the second question, *Can a woman be a Shariah court judge?*, does not use any Arabic religious or legal terms at all. So, as far as the lexical repertoire is concerned, it is less variable than the answer to the first question. The most interesting data are in terms of content (the arguments brought forward for why a woman cannot be a Shariah court judge) and the way these arguments are drawn up and presented.

A common trait of both LTE is that the RE uses metaphors and metaphorical language to a large extent and also uses quotations from the Qur'an in support of his theoretical statements or to convey some authoritative power. The Western, non-Muslim author of LTE (8) points out in her question (or statement) that the question concerning the equality of men and women really results from a linguistic problem, the fact that the adjective *equal* according to the OED has several meanings, one of them being 'identical in value' and another 'having the same rights and privileges'.

So the LTE author states that the question, whether Islam considers men and women equal, has to be answered according to the meaning of equal. The RE does not really take up this linguistic question but rather argues the matter itself.

The most interesting linguistic contrast of the two answers is the way the LTE authors are addressed (or not addressed) by the RE and the way he refers to them in the text. Whether or not the RE addresses the LTE author by the second person pronoun *you* obviously depends on the question asked in an LTE. If, for instance, the LTE author asks, *Am I as a wife responsible to settle my husband's accounts?*, then it is to be expected that the RE answers by *You should do ...* . If, as in LTE (9), a woman asks a general information question, *Why can't women lead congregational prayers* ... ?, the RE (as in most LTE, as far as I can see) answers without direct address of the LTE author, just by referring to the question asked by a certain person.

A very exceptional or even unique fact is that in LTE (8) asked probably by a non-Muslim Western female LTE author, the RE refers to her by title and last name, while giving an evaluation of her LTE. He starts his answer with, *The case is certainly well argued by Mrs McIntyre, but I still disagree with her*. I have not seen any other LTE in a Saudi newspaper in which the RE in his answer refers to the LTE author by her or his family name. This violates a text type-specific rule, so-to-speak, excluding reference by proper names to LTE authors. The reason here is obviously to express some distance to the LTE author as (probably) a non-Muslim, that is, a non-ingroup (or outgroup) member.

Another feature of formal properties in both answers is a rather frequent use of parallelisms and other rhetorical figures. Also the fact that certain attributes, appositions and other expressions are repeated in almost stereotypic fashion, such as in LTE (9) in answering the question why a woman cannot be a Shariah court judge, *Also a woman, despite her high degree of intellect and the high degree of her knowledge ... ; irrespective of their mother's high intellect ...* . The editor is putting much emphasis on the fact that women have a high intellect, so that one could almost be tempted to assume that some part of the readership would tend to assume the opposite.

Another remarkable linguistic fact is that the RE in LTE (9) inserts some rhetorical or metacommunicative comment on the fact that a *why*-question 'could be asked' (*One may ask, why?*), this being the very question that is being asked in the LTE and he is answering. So in introducing the issue whether a woman can be a Shariah court judge, stating the fact that women are not allowed to do so, the RE seems to give an indication that 'normally' such a question is not asked or that the answer to it is self-evident anyway.

An interesting 'marked' linguistic feature of LTE (8) is the use of an expression to a Western non-Muslim woman relativizing the RE's judgement, which is not very frequent in the LTE investigated nor indeed in the speech habits and style of the RE: ***To my mind**, equality cannot be* *administered unless ...* . Another lexical characteristic in LTE (9) is the expression *womenfolk* as some kind of an old (-fashioned), unusual term, addressing a totality of women, such as 'congregation of women praying'.

Even more specific seems to be the textpragmatic argument behaviour displayed in the RE's answers, in particular the differences on the descriptive level of **textual** or **textpragmatic** functions in the answers of the RE.

Several of the data mentioned noting differences in the component 'message form', in particular the lexicon, and 'message content' show interesting cultural differences between the two items. They contain a lot of culture-specific and religion-specific information on Islamic society and the position of women in it. Differences in the vocabulary, like the occurrence of the somewhat ancient term *womenfolk* (denoting the women's congregation during *salaah*) in LTE (9) directed towards a Muslim woman versus LTE (8) with no such expression and directed to a (likely) non-Muslim woman, are remarkable in their own right. Even more significant are the differences in the language and argument strategies used by the RE towards a Muslim woman from the East versus a probably non-Muslim woman from the West. Following, a brief interpretation of the argument structure of LTE (9) is given. Its aim is: (1) to show how the (male) RE answers the question of the Muslim woman from the East in LTE (9), what points he is making, and what strategies of persuasion and argument are employed; and an interpretation will be given (2) of what the differences between the two answers the RE gives to LTE (8) and LTE (9) are in these terms. The results can be summarized as follows:

1. Comparing the data of this corpus with these analysed for LTE by men (Kniffka, 1994, 2002), there is little difference in the content and argument strategies used by the RE in answering LTE by women and by men. It is safe to say that, generally speaking, the RE uses the same repertoire of communicative means and strategies to answer LTE by women and by men.
2. Not surprisingly, but nevertheless highly remarkable for cultural contrast and the role of women in it, there is a marked difference between the RE's answers to LTE (8) asked by a Western, probably non-Muslim woman and his answer to LTE (9) asked by an Eastern Muslim woman. It seems that, to put it in binary terms, the feature [+/− Muslim] is the most salient and prevalent 'distinctive feature' for the argument and the repertoire used by the RE. So it is this difference in the addressee component ([+ Muslim] woman from the East versus [− Muslim] woman from the West) that constitutes

the major difference expressed by the linguistic (semantic and pragmatic) means used in the RE's answers. At any rate, the component 'audience' (i.e., readership) and the impression the RE wants to create on it by his answers is to be accounted for.

In the following, some data to illustrate this are given. It should be added that the general perspective of the linguistic observer (i.e., my own perspective) is that of a Western non-Muslim male reader. Though I try to arrive at intersubjectively valid and reliable generalizations, it goes without saying that the results would be different if the linguistic observer were either a Muslim male from the East or a Muslim from the West, or a female from the West or from the East. Methodologically, it is necessary to have one specific receiver perspective to describe the difference in the sender's or addressor's component (the answers given by the RE). It is this variation which is in focus here, not a variation in the receiver's or addressee's perspective as such.

The first question in LTE (9) asked by a Muslim woman from the East is answered by the RE in quite 'simple and direct' terms. After enumerating the differences of various schools of thought in this regard, he states that all schools agree that a woman cannot lead a congregational prayer for men (same as a hermaphrodite cannot). If at all, she could lead only the *salaah* for womenfolk (Imam Malik's point of view). In addition, the argumentation simply states that, as far as reading the Qur'an and singing is concerned, a woman's voice is regarded as *awrah*, which would *immediately disqualify her and ... also evoke mixed emotions from the male worshippers*. A man's voice is not regarded as *awrah*, the RE continues, and that is the end of the argument. The reason, to which the *why*-question of the LTE author alludes, is simply given as the religious tradition in general and a particular attribute applying to the voice of women versus men. That is it.

It goes without saying, that this argument may be sufficient to a Muslim (woman), but hardly for a non-Muslim (man or woman).

The RE's answer to the question whether or not a woman can be judge of a Shariah court is even more interesting from a textpragmatic and argument point of view. The RE states the answer first: *A woman may not be a judge in a Shariah court nor is a woman allowed to bear testimony on certain matters in a court*. The RE then proceeds to pose the question (see above): *One may ask, why?* and also qualifies the answer to the question (in a semi-'performative' act): *The answer to this is very simple and very logical*. This is not really an argument at any rate, rather an abuse of argument. It is simply a claim (*It is very simple and very logical*). The RE does elaborate on a few practical reasons for this in the following text. He says that a judge must be well versed with everything that goes on outside, in the marketplace, the mosques and so on, and a judge would need to have discourse with other jurists in juridical matters, he would have to interview witnesses and so on. All this, factually correct, is forbidden for women in Saudi society: women are not allowed to sit with strange unrelated males *due to fear of temptation*, which the RE adds in parentheses.

As mentioned above, the RE takes pains to clarify that *in spite of women's high degree of intellect* and *high degree of knowledge*, she is not qualified anyway since she is *still governed by a natural instinct, namely her deep compassion bestowed on her by the creator and she can do nothing about it*. This quality, a woman's *deep compassion*, disqualifies her as a Shariah court judge, because *in most cases she may have her judgement clouded*. The RE continues by giving the example of the case in which the death penalty would have to be decreed as punishment: *In most cases this would be impossible for her to do because of her natural compassion, which will cloud her logic* (paraphrased). Needless to say, all these points that the RE makes in his answer are not valid explanations. He does not really give reasons for the fact that a 'woman's logic' will be clouded, he simply states that the degree of compassion women have (not men, which is obviously implied) has that effect.

In other words (and in simplified terms), the RE does *not* give any real explanation nor any reason why women have a natural compassion and men do not, why this natural compassion will cloud a woman's logic. He simply states or claims that this is so. He does not offer any empirical evidence to support this claim, but rather resorts to an example of everyday life, which has even less relevance, let alone a convincing argument in support of the afore-mentioned claims: The fact that a woman's natural compassion will cloud her logic *comes to the fore very strongly* where children are concerned: *Children know how to get to their mother (in spite of her high intellect), unlike the firm father. A baby crying for months during the night would not experience anything derogatory said by his or her mother. The father, however, would lose his cool after a day or two, because that is the way that Allah created him*. The RE continues, because he must go to work and *therefore* [sic] *his reasoning is very logical* [sic]. He does concede that the mother *also needs her sleep*, but she has been *strengthened by Allah with inner strength and with such compassion, which overrides every rule of logic*. This ends the argument reasoning in this answer. The RE continues by adding a verse from the Qur'an, in which is generally stated that some things are reserved for men and others for women. The reason that in cases of bearing testimony a woman is only allowed to do so in matters of trade and personal law, and two women would be considered equal to one man testifying, is explained by the fact that *womenfolk generally do not participate in such acts*. Yet it is difficult to see why two female witnesses would be considered equal to one male witness, if this really holds: if women are not participating in such acts, then two women would be as good as one. In some court cases, like for example adultery, women are not allowed at all as witnesses (but only four men) since the details to be reported to the judge would not allow for a woman witness, because she is not supposed to report on such *filth, because she is sacred*. The juridical problem involved in being able to present four witnesses in case of adultery would be a different matter, which cannot be discussed here.

To use the RE's own words, one would have to say, as a distant neutral observer, that 'his reasoning is not very logical', or not logical at all.

In conclusion, with due respect to the religious scholar who writes as the RE here: as a Western observer one has to say that this argument cannot be accepted as being logical in one way or any other. There is no real valid explanation for anything, but rather an enumeration of claims and statements which Muslims believe: A woman has a natural compassion and therefore would not be suitable as a judge, whereas a man does not have this natural compassion and therefore would have a clear (unclouded) logic, so he can be Shariah court judge.

The argumentation in the other LTE is different in method, attitude, speech style, and argument.

Several of the textpragmatic functions in the RE's answer to LTE (8) by a non-Muslim Western woman have been discussed already, such as the unique feature that he addresses the LTE author by her last name or complimenting the argument by this LTE author. The strategy in answering her question is to even surpass her, that is, to be even more 'generous' in the interpretation, implying that the semantic difference given by the two meanings of the word *equal* in English does not imply that women are equal in value only, but also are equal in rights and privileges according to the Islamic view. His first point is, abbreviated, that even if people get equal rights and chances, they cannot exercise them in the same way, so that considerable differences in physical and material characteristics are inevitable. As a special example from 'real life' (parallel to the answer to LTE (9)) the RE chooses education, where the law requires all children be given the same rights and privileges. However, in spite of these laws some children have special needs, such as dyslexic children. So if they are given special attention and more time on their tests and assignments, then it seems the law would favour them as against other children who do not get this special treatment. The RE rightly concludes that *equality cannot be administered unless these [dyslexic] children are given the facilities necessary to learn at their own pace.* The RE modestly puts a relativizing expression *to my mind* in front of this statement, which is certainly a special gesture towards a Western reader. The point of this argument, if I understand it correctly, is that people are equal and yet they are not equal, and some may need special treatment like in education. The exact parallel between this example from education and the point asked by the LTE author is not clarified any further by the RE, however, so that the impression remains (with a Western reader) that in fact he more or less confirms her point that they are equal in value, but not in terms of rights and privileges, yet claims that they are equal in both.

In the next paragraph the RE states verbatim that men and women are *equal in God's sight, both in value and in their rights and privileges.* He then enumerates the examples that former readers have cited, for example that women are not allowed to marry four husbands, that they cannot divorce at will, that they are disadvantaged in their inheritance, and that two female witnesses may serve in place of one male witness. He continues to say that this is a rather superficial argument which takes things out of context, and in support of this he quotes a verse from the Qur'an (surah 107), which says 'Woe to those who pray'. He points out (quite convincingly) that this verse of the Qur'an has to be taken in context with the following, which says: 'Woe to those, who pray, but are heedless of their prayers; who put on a show of piety but refuse to give even the smallest help to others.' It is certainly a valid point that one cannot take an expression out of context. But again it is not clear, how this example is actually suited to confirm his view on the statement that the LTE author wants to have clarified, what its argument potential could be to support his claim. He does not explain how it can be used as an argument in this case. More exactly, the point made by Mrs McIntyre – women are equal to men in value, but not in rights and privileges – does not necessarily result from an understanding or quotation of anything out of context, which the RE fails to explain in convincing detail. Rather he goes on stating that *such a hasty and faulty judgement* [sic] should be avoided, and *we should carefully consider these differences in the rights and privileges of men and women.*

The RE claims that to take adequate account of all this, one has to conclude that the differences in rights and privileges *do not have any bearing on the equality between them [men and women]. They are meant only to help each of them to fulfil the role assigned to them.* So, if I understand the point correctly, there are differences in rights and privileges between men and women, but these do not imply inequality of the two and they simply result from the fact that different roles are assigned to men and women which they are supposed to fulfil. These God-given different roles are not open to argument and explanation it is implied. This reminds one of a '*deus ex machina*' explanation in medieval texts. The concluding remark (not really applying to an explanation of the point made by the LTE author) is that men and women have the same duties to believe in God and *to worship Him in the same manner* and that *both will have the same reward for any act they do in fulfilment of their religious and community duties*, which all seem to sum up the assumption that because of that both must be equal.

The overall impression of the contrast between this answer of the RE towards a Western non-Muslim woman and his answer to an Eastern Muslim woman is that much less, or none, of the statements about women's place and role in society given in LTE (9) are given in LTE (8). A somewhat convincing example of the inequality of people in spite of equal opportunities is given in LTE (8), which is not, however; an appropriate point to show that women are equal both in value and in rights and privileges. In other words, it fails to be an adequate explanation of the RE's thesis and does not apply to the case given.

The other argument supplied by the RE in his answer to LTE (8) that one should not quote statements out of context is in itself very true, but does not help to support his claim or to answer the implied questions by the LTE author either, since he does not elaborate that Mrs McIntyre does in fact consider things out of context in coming to her conclusion. He fails to give sufficient evidence of why the LTE author is making *such a hasty and faulty judgement*. Again, in much more elegant and 'Westernized' rhetorical terms than in the answer to LTE (9), he still fails to rationalize the point made and to give an adequate explanation for the correctness of his own view. Same as in the answer given to a Muslim woman, the conclusion is derived from a general religious view which is not suited as an explanation, but gives a sufficient answer in some kind of a *petitio principii* which a strict Muslim would accept, a non-Muslim probably would not. Both, women and men, have the same religious duties and get the same rewards. So they are bound to be considered as equal in value and in rights and privileges?

The answers of the RE to different LTE try to meet the expectations of the LTE authors (and readers) from a different background in the best possible fashion. The main contrastive aspect in this corpus is the contrast of the answers to a Muslim LTE author from the East versus a non-Muslim LTE author from the West. One can say that the method of argumentation, the use of textpragmatic means to get his point across, are different, the ultimate aim and the basic argument structure in both cases being much the same.

It goes without saying that the RE, within the bounds of his job, his religious conviction, and trying to do optimum justice to both authors/addressees, fails to really give a full-fledged explanatory answer (at least by Western standards) to the questions asked. Eastern Muslim readers may find his explanations very convincing though. At any rate, this behaviour towards women by a male Religious Editor reveals culture-specific and religious ingredients: how women are treated and addressed in a religious context, how their arguments are handled by (male) experts in an authentic culture-specific fashion.

Conclusion

Men's and women's LTE look very much alike in the culture and media investigated. Topics addressed, message form, message content, text type characteristics and other features are very similar if not the same in both. Men and women both write about roughly the same topics or, more exactly, the LTE printed in a newspaper do not reveal any major differences in that respect for men and for women. This can mean that as a '*genus commune*' it shows in fact a common trait of the work of the Religious Editor responsible for which LTE are printed in which form, rather than of the LTE authors.

Men's and women's LTE published in Saudi Arabian newspapers do show some significant differences in how the LTE authors are 'identified' or referred to by names in the paper. Men are the 'unmarked' class of authors, for which abbreviations (initials) or first names are sufficient. Women are identified by adding *(Mrs)* to the name. In addition to the feature just mentioned, some text-internal and text-external data can be stated as reflecting and expressing/carrying culture-specific features of Saudi Arabian culture; for example, a 'correlation' of a particular topic and the withholding of the name and address of an LTE author by the Religious Editor.

LTE represent a structurally unique text type (in Saudi Arabian and in other daily newspapers) of mass-media communication, in that at least three structural criteria are involved:

1. LTE represent a 'person-to-person' dyadic interaction, between (many) LTE authors and (one or more) 'Religious Editor(s)' within a large and complex non-person-to-person communication dyad of the newspaper's staff and its readership.

2. Some receivers (readers) function as senders (LTE authors) in/for the newspapers which gives them a unique status.
3. Both parts of the message form of the text type LTE, or this particular sub-text type of LTE consisting of questions and answers appear in print in the newspaper, that is, the original 'person-to-person' interaction appears in public, which may trigger further interactions and recruit other newspaper readers as LTE authors.

This is the only text type of mass-media communication via newspapers (possibly with the exception of personal ads) where the 'voice of the reader' can be heard/read, which makes it a valuable source of culture-specific data of critical importance.

This is the only place in Saudi Arabian culture in which the 'voice of women' can be heard/read, be it 'authentic' or modified by the (male) Religious Editor to varying degrees in various fashions. In this sense, LTE by women in the 1980s and 1990s are certainly one of the most important sources of data for identifying the culture-specific values held in Saudi Arabia by and about women in this period.

Notes

1 See Hymes, 1962, pp. 13–53; Gumperz and Hymes (eds), 1972; Hymes, 1972, pp. 35–71; Baumann and Sherzer (eds), 1974; Baumann and Sherzer, 1975, pp. 95–119; Saville-Troike, 1982, 1989. One such adaptation of Gumperz and Hymes, 1972 for the analysis of newspaper communication has been described elsewhere (Kniffka, 1980, pp. 22–39).

2 A more detailed analysis of the semiotic status of the name reference is given elsewhere (Kniffka, 2005).

3 FN = First Name; LN = Last Name; CN = City's Name.

References

Baumann, R., and Sherzer, J. (Eds). (1974). *Explorations in the ethnography of speaking*. London: Cambridge University Press.

Baumann, R., and Sherzer, J. (1975). The ethnography of speaking. In: *Annual Review of Anthropology* 4, 95–119.

Bell, A. (1991). *The language of news media*. Oxford and Cambridge, MA: Blackwell.

Bucher, H.-J. (1986). *Pressekommunikation. Grundstrukturen einer öffentlichen form der kommunikation aus linguistischer sicht*. Tübingen: Max Niemeyer Verlag.

Herzog, C., Motika, R., and Pistor-Hatam, A. (1995). *Presse und offentlichkeit im Nahen Osten*. Heidelberg: Heidelberger Orientverlag.

Hymes, D. (1962). The ethnography of speaking. In: T. Gladwin and W. C. Sturtevant (Eds), *Anthropology and human behavior*. Washington, DC: Anthropological Society of Washington (pp. 13–55).

Hymes, D. (1972). Models for the interaction of language and social life. In: Gumperz and Hymes (Eds), *Direction* (pp. 35–71).

Gumperz, J. J., and Hymes, D. (Eds). (1972). *Directions in sociolinguistics: The ethnography of communication.* New York: Holt, Rinehart & Winston.

Kniffka, H. (1980). *Soziolinguistik und empirische textanalyse. Schlagzeilen- und leadformulierung in amerikanischen tageszeitungen.* Habilitationsschrift Köln 1980. Tübingen: Max Niemeyer Verlag (= Linguistische Arbeiten Nr. 94).

Kniffka, H. (1994). Letters to the editor across cultures. In: *Intercultural communication. Proceedings of the 17th International LAUD-Symposium,* Duisburg, 23–27 March 1992. H. Pürschel et al. (Eds). Frankfurt am Maine: Peter Lang Verlag (pp. 381–409).

Kniffka, H. (2001). Dialogical genres of newspaper communication across cultures. Letters to the editor in English Saudi Arabian daily newspapers. In: U. Fix, St Habscheid, and J. Klein (Eds), *Zur kulturspezifik von textsorten.* Tübingen: Stauffenberg Verlag (pp. 255–89).

Kniffka, H. (2002). Sprach- und kulturkontakt 'across the fence(s)'. Linguistische perspektiven. In: E. Apeltauer (Ed.), *Interkulturelle kommunikation: Deutschland – Skandinavien – Großbritannien.* Tübingen: Gunter Narr Verlag (pp. 7–30).

Kniffka, H. (2005). *Degrees of 'anonymization' of senders' names in letters to the editor in daily newspapers.*

Labov, W. (1970). The study of language in its social context. In: *Studium Generale* 23(1), 30–87.

Labov, W. (1972). *Sociolinguistic patterns.* Philadelphia: University of Pennsylvania Press.

Saville-Troike, M. (1982, 1989). *The ethnography of communication. An introduction.* Oxford: Basil Blackwell.

7

A Cyber-Parish: Gendered Identity Construction in an On-Line Episcopal Community

Sage Graham

Technology has historically had a tremendous impact on religion, from the invention of the printing press (which allowed the Bible to be read by mass populations) to the Evangelical Christian movement (which used radio and television to spread the word and preach to mass audiences). Just as religion has historically had an enormous impact on society, so also has the recent advent of the Internet changed our social and organizational frameworks. In the twenty-first century, these two powerful social forces have come together with the advent of online religious groups ranging from Usenet discussion groups on religious topics to virtual preaching and religious services. Participants in various types of religious groups can request prayers, inquire about others' beliefs and/or doctrine, and even receive sacraments and communion via virtual blessings of bread and wine. The ability to worship and/or form religious communities online has tremendous implications for both participants and organized religion itself.

While some researchers have begun to explore computer-mediated communication (CMC) and religion (e.g., Schroeder, Heather and Lee, 1998), and several linguistic studies address gender in a computer-mediated setting (Herring, 1994, 1996a, 1996b; Hall, 1996), there has been no research to my knowledge on how religious group identity and community are influenced by gender in a computer-mediated setting. Within the context of religious communities, an understanding of gender as it intersects with computer-mediated communication is pivotal for three reasons:

1. Previous research on CMC online (Herring, Johnson and DiBennedetto, 1992) has indicated that cyberspace is just as

male-dominated as 'real-world' settings; women are frequently marginalized in all types of e-settings (academic lists, in the case of Herring). This is not the case on ChurchList, the email discussion list which is the focus of this study, however. In this setting, women occupy a pivotal role in clarifying and maintaining the group identity and community. It is important to research the strategies these women employ to participate in this 'cyber-parish' community, since their active participation runs counter to both the 'real-world' church and the Internet (both traditionally spaces that have been dominated by men).

2. Within the Anglican Church (the denomination of ChurchList), the role(s) of women within the Church have been hotly debated since the 1970s. If cyber-parishes offer women a means of active participation that they are denied in 'real-world' church settings, then they may choose cyber-church over (marginalized) participation in 'real-world' church.
3. As religious groups depend more and more on online interaction as a means of disseminating information and strengthening church ties, the advent of computer church communities is at once potentially empowering and (potentially) destructive to traditional 'real-world' church parishes.

It is the goal of this study to examine gender roles in an online Episcopal church group (cyberparish) to explore the ways that women and men participate in this cyber-community – ultimately trying to determine: (1) if women are marginalized in this setting; and (2) what communicative and linguistic strategies they employ to claim authority within the group. I will employ interactional sociolinguistic analysis of empirical data, exploring: (1) the frequency of participation among male and female listmembers within the community; as well as (2) strategies employed by community members to shape the group identity. Through this analysis, I will show that women play an active role in the formation and maintenance of the e-community, and, possibly, through their participation it is likely that they also strengthen their roles (and empowerment) within the 'real-world' church.

Previous research

Gender and religion

> Radical [female clergy] are, but radical they must not appear to be.
> (Lawless 1996b, p. 405)

To date, there has been limited research focusing on the relationship between gender and religion as it is manifested in language use and linguistic strategies. Research on gender and religion does state, however, that formal religion has traditionally been male-centred. As Clark and Richardson (1977) note, 'the Western Religious tradition ... is unremittingly patriarchal'. They go on to explain that 'progress towards a more enlightened view of women is ... real, but it is also painfully slow and it proceeds against a constant patriarchal opposition which has become more outspoken with the passage of time' (p. 1). Evidence of the resistance to women occupying positions of power within this patriarchal system can be seen in Lawless's *Holy women, wholly women* (1993), which examines life stories of female clergy members. Lawless explains that women face particular challenges as they take on authority roles in religious settings, and notes that 'Quietly, women are invoking a new religious era' (p. xiii). She expands on this idea of female revolution within religious structures in both 'Images of God in Christian women's sermons: Finding God' (1996a) and *Women preaching revolution* (1996b), both of which examine differences between female sermons and more traditional male sermon models, noting the inadequacy of sermon-training for female clergy members who employ different tactics than their male counterparts. She goes on to explore the differences in approach between male and female clergy, stating that:

> Women in the clergy create a space where shared power and authority and collaborative learning about things spiritual and religious can happen through free and unrestricted dialogue and respect for difference. Women in the ministry, in the pulpit, preaching, leading prayers, healing, conferencing together are creating the framework for [a] new dialogic paradigm.
>
> (p. 168)

Despite these hints at change, women in the Anglican and Episcopal Church (the Church which is the focus of this study) have not necessarily been embraced in their attempts to occupy leadership roles (and thereby take positions of power) within the Church hierarchy. Although, as of the 1998 Lambeth convention,[1] most Anglican provinces ordained women and seven provinces allowed females to become bishops,[2] the subject of women as leaders in the church is still controversial and is still debated within the Church today. As recently as 1999, Episcopal bishop Jane Dixon was denied entry into a church in her diocese because the conservative parishioners did not approve of females being ordained as clergy. This tension regarding the power of women within the structure of the Church begs the question of how women achieve power (in what settings), what (linguistic) strategies they use to maintain their positions of power/authority, and what implications this has for women's place in the world-wide Church. It is one goal of this study to examine these questions, taking the parish structure of ChurchList and examining the ways that female participants: (1) take an active role in the formation and maintenance of the cyber-community; and (2) thereby occupy places of power within the cyberparish structure.

Conflict as a means of establishing power

In addition to examining the dynamics of women within religion, this study also draws on previous research that examines the intersection of conflict and gender. Conflict, as a social phenomenon, can be an important tool in negotiating (group) identity, and can therefore be used as an analytical focus for examining gendered power roles in establishing and maintaining that group identity. Although there has been a great deal of linguistic research on conflict in face-to-face settings (Goodwin, M. H. and Goodwin, C., 1987; Corsaro and Rizzo, 1990; Eder, 1990; Goodwin, 1990a, 1990b; Grimshaw, 1990; Philips, 1990; Tannen 1990, 1994, 1998), studies which examine conflict in computer-mediated contexts are much more limited. Herring (1994, 1996a, 1996b) and Herring, Johnson and DiBennedetto (1995) show that, consistent with Tannen's (1990, 1994, 1998) findings, females who participate in e-conflicts cushion their disagreements through the use of affiliative comments and posing questions rather than making assertions. Male posters, on the other hand, use a more adversarial style which establishes/reinforces hierarchies through promoting their own views while knocking down the views of others. Conflict is, nevertheless, an important focus for study in examining gender and religion because, through conflict, interactants can both establish camaraderie and negotiate status and power within their relationships. The ability to use conflict as a means for establishing power and authority is particularly important within the cyber-parish of ChurchList; if women can assert their power by actively negotiating the resolution of conflict, then they are also empowered to enforce their active presence as participants within the community (and might then extend that power to 'real-world' church settings). Within this group, moreover, the ability to use conflict to establish community ties is also pivotal; establishing connections is a central idea of church groups. Females' active participation in the negotiation of conflict in the

ChurchList setting, then, has two results: (1) it allows women to take an active (and powerful) role in forming and reforming expectations for group identity; and (2) it provides a means for the whole group to reinforce their solidarity (male and female) as group members (while simultaneously reinforcing the authoritative role of women in establishing that solidarity).

Data and methodology

In order to more fully understand the interaction between gender, computer-mediated communication and religion, I take an ethnographic, interactional sociolinguistic approach to analysing a series of messages sent to ChurchList, a self-organized, international discussion list that exists to provide a forum for discussion of issues that affect the Episcopal and Anglican Churches. In order to determine the impact of gender on participation, I analysed all messages sent to the list between 28 December 1997 and 7 January 1998 (n = 2135). As I show in Graham (2003), the messages sent to the group during this time period are representative of interactions on the list throughout the rest of the year. I chose to more closely examine this particular time frame, however, because it was during this time that a large conflict arose which threatened the survival of the list. It is my claim that, through the negotiation of large-scale conflicts like the one that is the focus of this study, female participants take an active role in the negotiation of the group identity. Building on Harré and van Langenhove's (1999) theories of positioning, I therefore examine the ways in which male and female subscribers position themselves (particularly with regard to power and authority) in relation to the cyberparish.

There are several important features of ChurchList that are worth mentioning here since they affect the negotiation of group identities and the norms for interaction among listmembers (both male and female):

1. ChurchList is both 'open' (meaning that anyone can subscribe to the list at any time) and unmoderated (meaning that all posts are distributed directly to subscribers without being 'approved' by a moderator). In this setting, then, listmembers themselves determine and control the development, content and flow of conversational topics and take an active role in the formation of group goals and norms of interaction.
2. Though the volume of daily messages and the number of individual subscribers is quite high, a majority of the posts are actually sent by a

small percentage of the total subscriber population. This 'core' group has a tremendous impact on communications and the community structure; they socialize newcomers and, when conflicts arise, play a large role in the resolution of the conflicts and the subsequent negotiation of the group identity.

3. A final characteristic of this list that merits discussion is the community orientation and 'feel' of the list. Traditional anthropological definitions of community have been dependent on geographic location (Hymes, 1974), but as researchers (Hamilton, 1998; McWilliams, 2001) have begun to document, it is very common to form a 'community' without geographic proximity as the defining factor. It is not unusual, therefore, despite the fact that listmembers are spread all over the world and may not encounter one another face-to-face, that ChurchList is, in fact, a very closely knit community.

In order to examine gendered communication patterns within this dynamic cyberparish, I first performed a quantitative analysis of the posts to the list. This analysis (following Herring, 1996a) is driven by two primary questions: (1) what is the average rate of participation of males and females in this cyber-community?; and (2) are there differences in the kinds of messages that males versus females post? Using the 2135 messages sent to ChurchList between 29 December 1997 and 7 January 1998, I correlate gender with the total number of posts in an attempt to determine variation in the amount of participation between male and female group members.[3]

In addition to examining overall participation by males and females, I also examine participation rates within the group of 'core' listmembers mentioned above. Since core (versus non-core) members participate more actively and more frequently within the community, I track male and female participation inside the 'core' group to determine the impact of gender within this sub-set group (which, because of their longevity and long-term commitment to the e-community has a great deal of power to influence norms of interaction and behaviour). For this study, I identified members of the core group based on posting statistics that were compiled (voluntarily) by one listmember and sent to the list each month. In January, this same listmember compiled and sent posting statistics for the previous year. For my purposes, I identify core members as those who post over 100 messages to the list over the course of a calendar year. I believe that this participation rate of 100 or more messages per year indicates a stable list presence and a level of commitment to the group that becomes pivotal in establishing norms of community interaction and socialization of newcomers. In the following analysis, core membership is also important because participation (number of messages sent to the list) is greater (for both male and female listmembers) among the core group.

Finally, I perform a qualitative analysis of messages posted during a conflict within the group. The primary goal of this qualitative analysis is to evaluate types of female participation in conflict negotiation sequences, since I would argue that conflict is one primary way that group identity is negotiated. Those listmembers who actively participate in the negotiation of the group identity and expectations are therefore empowered and active members of the e-community. Participation by females in this context, then, is a valid indicator of female authority and power within the group.

Analysis

Ratio of male to female overall participation

As I mentioned above, previous linguistic research on computer-mediated communication has shown that males tend to post more messages to online discussion lists on a regular basis than females (Herring, 1996a). Furthermore, this same study also indicates that when female participation rates rise above approximately 30 per cent (which occurs during the conflicts that Herring examines), male community members begin to try to end the discussion by criticizing the females for talking too much and taking the discussion too far. In the e-community that is the focus of this study, however, females post close to 40 per cent of the total messages on a regular basis. Moreover, when conflicts arise, female participation stays at or above 40 per cent; females not only participate this much in 'regular' interactions, they also play an active role in conflict negotiation which empowers them to influence and shape the group identity and norms for interaction. Table 7.1 indicates the overall ratio of male to female participation.

As Table 7.1 shows, the percentage of female participation on ChurchList is higher than previous research on other discussion lists has

Table 7.1 Overall male and female participation rates

	Core	Non-Core	Total
Male	953	194	1278 (60%)
Female	663	325	857 (40%)

indicated. This rate of participation, however, is not adequate to demonstrate the complexity of male and female roles within the cyber-community. In order to more fully break down the complex participation rates of male and female listmembers, then, it is necessary to more thoroughly examine the relationship of gender to core versus non-core member status.

Male versus female participation within the core group

In my analysis of this e-community, I have identified two groups of participants: Core and Non-Core. Although (as previously mentioned) the subscriber rate of ChurchList is quite high, among the subscribers is a much smaller 'core' group which posts the majority (approximately 75 per cent) of the messages to the list. Through active (and consistent) long-term participation, this core group plays a key role in establishing and maintaining the norms for interaction within the list community. As previously mentioned, while Herring (1996) found that female participation stayed below 30 per cent on the discussion lists she examined, within ChurchList female participation is regularly closer to 40 per cent. It is not enough to simply calculate the number of posts sent to the e-community, however, since this type of calculation does not take into account the fact that there is variation in the number of subscribers who are male versus female. A single list, for example, might have a three to one ratio of male to female subscribers. On this same list, one might note that male listmembers post three times the number of messages than females. Despite the fact that the participation rate of males is 75 per cent (of the total posts), however, this number does not take into account the disparity in the number of male to female community members. In a situation like this, it is conceivable that female community members might in fact post MORE messages than the males, but, because of their fewer numbers, still do not post a large percentage of the overall number of messages. To track the rate of female postings within the core group, therefore, I calculated the total number of male and female core group members and divided that by the total number of messages sent by the male and female core groups. This yields the average number of posts for each group (an important consideration since it takes into account the fact that there are fewer female listmembers in addition to there being fewer messages posted by females). Table 7.2 shows the ratio of participation among male and female core listmembers.

In this case, the average number of messages posted by female core listmembers is actually greater than the average number of messages sent by male community members. Although the number of female core

Table 7.2 Male and female participation rates within the core group

	Number of Female and Male Core Listmembers	Total Number of Messages Sent by Core Females/Males	Avg. Number of Messages Sent by Each Core Listmember
Females	45	663	6.78
Males	60	953	6.29

listmembers is smaller than that of males, the average number of messages sent by both males and females within the core group indicates that individually, females in this core group participate as much as males. This fact sets this list community apart from the ones examined in most other linguistic studies (Herring, 1996a) and indicates a level of participation and empowerment that females on other types of lists may not experience.

Male and female participation in conflict as a marker of empowerment within the cyberparish

Participation is only one component of establishing authority within this community, however. In order to gain a more comprehensive understanding of the dynamics of an e-community, one must not only examine the number of posts, but also the content, purpose and quality of those posts. As Tannen (1990) notes, men and women tend to employ different strategies and work towards different goals in interactive settings; women work to establish *rapport*, while men tend to *report*. As Herring (1996a) further explains, Tannen's schema for the types of messages produced by males versus females does not necessarily apply in a computer-mediated context. In Herring's examination of two academic e-mail lists, she found that communication was more often governed by the norms of the list itself (some lists taking adversarial posts as the norm) rather than gender. The concept of adversarial interaction being unusual for females is an important one for the purposes of this study. When conflicts arise on ChurchList, messages sent by female community members are frequently just as adversarial as those of male contributors. The difference between male and female participation tends to emerge in the type and goal of the posts sent by females versus males. I will now examine strategies employed in specific messages to identify communicative patterns employed by males versus females.

The conflict I examine here arose when a (male) listmember posted a request for prayers (listed in Example 1).

Example 1

1 Subject: Miserable – Pray for Me

4 I feel miserable and lost. I wish I were dead. I am unemployed and so is
5 Carl. We are living on savings that will eventually run out. I am getting
6 to an age where people won't want to hire me anyway. I have just been
7 turned down for the church administrator job I wanted. All this is making
8 me so depressed I have a hard time presenting myself well to a prospective
9 employed.[4] I have just finished temping in a place I really liked. I can't
10 practice the profession I am trained for. In fact, I am having a hard time
11 getting licensed to do it as a volunteer. I can't do what I have the gifts
12 for, preaching and teaching, celebrating liturgy and the spiritual care of
13 people. Carl says he thinks of me more as a friend or parent than a boy
14 friend or lover and is seeking someone else. And I know that I can't give
15 him what he wants and needs sexually anyway. He continues to
16 expect me to support him while he does this. But if I show an
17 interest in someone he
18 gets very jealous and angry at me. In addition, he is so emotionally
19 afflicted that he can't work, and is totally dependent on me financially
20 and emotionally. I feel so lost and alone and unsupported. What is going to
21 happen to me. Will I end up on the streets or what?
22 Brad[5] <email@email.com>

The first response to this prayer request was also sent by another core listmember: Jane D. Her message is included in Example 2:

Example 2

1 Subject: Re: Miserable – Pray for Me

4 Brad wrote:

5 >[6] In fact, I am having a hard time
6 > getting licensed to do it as a volunteer. I can't do what I have the gifts
7 > for, preaching and teaching, celebrating liturgy and the spiritual care of
8 > people. Carl says he thinks of me more as a friend or parent than a boy
9 > friend or lover and is seeking someone else. And I know that I can't give
10 > him what he wants and needs sexually anyway. He continues to expect
11 > me to support him while he does this. But if I show an interest in
12 > someone he gets very jealous and angry at me. In addition, he is so
13 > emotionally afflicted that he can't work, and is totally dependent on me
14 > financially and emotionally. I feel so lost and alone and unsupported.
15 > What is going to happen to me. Will I end up on the streets or what?

And people wonder at the gift of scripture's dictate to confine sexuality
to heterosexual marriage. Perhaps scripture goes this way as an
attempt to reduce the overwhelming confusion of boundaries listed here.
Perhaps scripture is not the arbitrary authoritarian abuse it is seen as
today, but an expression of understanding of what people have to go
through in their lives
Jane

Jane's response to Brad's request began an ever-multiplying web of responses which called the group expectations into question and ultimately threatened the e-community itself. Like the other large-scale conflict which I witnessed during my time as a ChurchList subscriber, the progression of the conflict began with an examination of individuals' roles within the e-community and then expanded to provide a forum for the (re)negotiation of the group identity and expectations.

One important factor here is the impact of gender in conflict participation (and thereby group identity negotiation). In the next section, I examine specific messages sent to the list during the conflict in an attempt to determine what, if any, differences exist in messages posted by males and females.

Of the messages that contribute to the negotiation of group identity, 113 address norms and expectations for list behaviour (i.e., whether Jane's response to Brad and the subsequent attacks on Jane by other listmembers constitute appropriate and/or acceptable behaviour). Of these 113 messages, 61 (54 per cent) were posted by female core listmembers while the remaining 52 (46 per cent) were posted by male core listmembers.

In the first stage of the conflict, which begins immediately after Jane's response to Brad (listed above), listmembers attack Jane for behaving in an inappropriate way in responding to a listmember in pain. These posts range from offering the suggestion that Jane's response was inappropriate or misplaced to vehement attacks which criticize Jane for being insensitive (and worse) in her response to Brad's initial message. Some examples of these messages are included in Examples 3 through 6:

Example 3

[...]
Jane, I'm writing to Brad privately, but to you I say publicly: **this
is crueler than I can imagine** anything could be from someone
claiming to be a Christian ...

21 **surely**
22 **your call from God is not to inflict suffering on others**. Do you not
23 know that when you write words, when you speak and act, you hold other
24 people's hearts and souls in your hands? It is a great responsibility,
25 and one I beseech you to take more seriously.
26 Chelsea
[emphasis mine]

Example 4

[...]
33 My God in heaven ... a dear listbrother is hurting ... hurting in a way I
34 wouldn't wish on my worst enemy ... and he trusts us so much that
35 he bravely writes us a heart-wrenching note about his troubles.
36 What happens? **He gets kicked in the teeth by a listmate in one**
37 **of the most ugly and vindictive pieces of dreck I've ever read**
38 **on ChurchList. I am horrified and damned near in tears. I**
39 **don't know what kind of mind games Jane likes to play, but**
40 **when a person expresses the wish that they were dead ANY**
41 **reasonable and prudent person wouldn't write the above**
42 **quoted ... well, umm, feces.**
[...]
48 Respectfully and humbly,
49 Jarvis A. Edwards
[emphasis mine]

Example 5

[...]

4 dear people of Church-L, especially Jane D.,
5 Jane, you responded to my friend Brad's heartfelt plea for prayer and
6 understanding with this:

[...]

12 > Perhaps scripture is not the arbitrary authoritarian abuse it is seen as
13 > today, but an expression of understanding of what people have to go
14 > through in their lives.
15 >
16 > Jane
17 ... and *perhaps*, **Jane, you are a nasty piece of work so full of**

18 **her own agenda and looniness that you cannot see much of**
19 **anything at all ...**
20 Consider yourself filtered, JD – for ever ... by this address. Life is
21 just too short to have to listen to anything else you might have to say.
[...]
32 Dora
[emphasis mine]

These examples are only a subset of the total responses to Jane's reply to Brad; these posts range from expressions of simple disagreement with Jane, to direct attacks on her behaviour (Examples 3, 4 and 5). In these messages, Jane is called cruel for inflicting suffering on others (Example 3), is accused of 'kicking a listmate in the teeth' and sending 'faeces' in her messages (Example 4), and called 'a nasty piece of work' (Example 5). These criticisms of Jane are directly linked to her action of criticizing Brad after he requested prayers and support. By criticizing Jane's behaviour in this way, these listmembers address a code of conduct in force within the community. By criticizing Jane's behaviour so directly, they make it known what type of behaviour is acceptable (and what type of behaviour is not).

It is also important to note that Jane is not the only one criticized for her list-behaviour. After Dora posts her response to Jane (Example 5), she also becomes the object of criticism. In many cases, for example, listmembers post messages which chastise Jane's attackers (as well as Jane herself) – not because of the views the messages espouse, but because the form of the disagreement is interpreted as inappropriate (just as Jane's original response to Brad was viewed as inappropriate by many listmembers). Some samples of these second-tier critical messages (which focus on the (in)appropriateness of Dora's actions) are included in Examples 7 and 8:

Example 7

2 Subject: The Filtering Game

[...]
31 **... if I were to put on a filter, I'd do it quietly without any**
32 **announcement. And if I felt I needed to let the person know that I**
33 **wasn't going to respond, I'd do it by a private e-mail note, not shout**
34 **it over the PA system.**
[...]
39 Peace, Joel
[emphasis mine]

Example 8

2 Subject: The Filtering Game
[...]
12 **Joel, I disagree with you on this one. At least in this particular case.**
13 **Sometimes we say we're filtering publicly to slap someone down (I**
14 **know I'm guilty of this), and that's not very nice. But sometimes it**
15 **seems to me**
16 **there's a clear need to stand in solidarity with someone who's hurting** and
17 who has been cruelly attacked. That's clearly what was happening here.
[...]
27 I appreciate your comment and respect your feelings. But **I stand**
28 **with Dora**, here.
29 Best,
30 Ashlie
[emphasis mine]

This series of messages was followed by a message from Dora, apologizing for the *way* she chastised Jane (although not apologizing for the content and criticism of her message). Dora's apology is followed by another series of messages justifying Dora's actions and thanking her for taking an active role in (publicly) educating Jane on appropriate list-behaviour. An example of one of these messages is included in Example 9:

Example 9

(To Dora, in response to Dora's explanation/apology for her message to Jane (shown in Example 5, above))

4 I don't think you have much to apologize for. Sometimes right is just
5 right, and wrong is just wrong, and **it's necessary for the health of a**
6 **community for someone to get up and say so.**
7 ... There's no excuse for kicking someone when they're down,
[...]
12 From a woman with decided opinions,
13 Margaret C.
[emphasis mine]

All of these messages demonstrate expectations about what is (or should be) appropriate behaviour within this community. Not only is Jane's behaviour criticized, but also Dora's response to Jane (Example 7). Dora is called to task for being so public with her ire towards Jane, but is also supported for her actions by other listmembers (Examples 8 and 9). This ongoing dialogue about what constitutes acceptable behaviour within the list community is not only important for the negotiation of the group identity, it is also significant because female listmembers contribute 54 per cent of the messages through which this negotiation takes place. Moreover, rather than being told to stop posting on the subject (as the females in Herring's 1996a study were), the female participants in this setting are thanked and validated for their role in reinforcing the expectations of the community by other listmembers (Example 9, lines 4–6). In this way, the posters of these messages (the majority of whom are female) make known their expectations for acceptable list-behaviour and thereby contribute to shaping the norms and expectations of the group. In this sequence of messages, then, the female listmembers: (1) take an active role in determining and negotiating what expectations exist for behaviour within the community; and (2) are validated in taking on that role by other community members.

Conclusion

As this analysis has shown, on ChurchList, females occupy positions of greater power on the list than has been evident in previous studies (both in terms of their rate of participation and in terms of their power to shape group identities and expectations). In this e-community, females play a greater role (than in other cyber-communities) in defining appropriate behaviour and negotiating the norms and expectations of the cyber-parish. Contrary to Herring's (1996a) findings, on this list females play an active role in shaping interactional expectations within the e-community. Although this study gives valuable insights regarding the influence of females within this e-community, the reasons are too complex to be determined from this case-study. Further research on other 'cyber-parishes' is needed to explore whether ChurchList is representative of other church-oriented communities (as opposed to the academically focused lists that Herring examined) or whether this particular list-community is unusual in the rate of male and female participation.

Further research on this topic is also needed because, not only will further research provide insights into on-line interactional patterns and strategies, it will also give further clues as to how empowerment in an on-line setting impacts the structure of religious communities in the 'real-world'. If females are empowered in on-line settings to contribute actively to the identity of the group as a whole, they may feel empowered to take a more active role in their 'real-world' parishes. If, on the other hand, the participation of women in the Church hierarchy continues to be controversial, women may be tempted to turn to e-religion, choosing empowerment and (more) active participation in a cyber-parish over controversy and marginalization in a 'real-world' parish.

Notes

1 This world-wide convention is held once every ten years so that church leaders from around the world can discuss issues affecting the Church and vote on proposed changes to doctrine.
2 The term *Anglican* includes the world-wide Anglican Church, the Church of England, and the Episcopal Church of the United States (ECUSA).
3 One methodological note is in order regarding my analysis of posts as being sent to the list by males versus females. As I note in Graham (2003), identifying the gender of the poster in an email context can be difficult, since males and females frequently share email accounts and/or since names attached to email accounts may or may not be gender-identifiable. In order to address this difficulty, I drew on my knowledge as a participant observer as well as analysing individual message signatures, photographs of individual members posted to a ChurchList web site, and pronoun usage within individual messages to determine the gender of each message sender.
4 This typographical error was present in the original message.
5 All names which appear in this study are pseudonyms.
6 A '>' symbol at the beginning of a line of an email message indicates material that is quoted from a previous message. In this case, Jane hit the 'Reply' button when writing her response to Brad and his message is therefore reprinted above her response. This type of quoting within email messages is particularly important on a high volume list like ChurchList, since it provides necessary contextualization information to help readers understand how to interpret individual messages.

References

Clark, E., and Richardson, H. (1977). *Women and religion: A feminist sourcebook of Christian thought*. New York: Harper Collins.

Corsaro, W. A., and Rizzo, T. A. (1990) Disputes in the peer culture of American and Italian nursery-school children. In: A. Grimshaw (Ed.), *Conflict talk: Sociolinguistic investigations of arguments in conversations*. Cambridge: Cambridge University Press (pp. 21–66).

Davies, B., and Harré, R. (1990). Positioning: The discursive production of selves. In: *Journal for the Theory of Social Behaviour* 20, 43–63.

Eder, D. (1990). Serious and playful disputes: Variation in conflict talk among female adolescents. In: A. Grimshaw (Ed.), *Conflict talk*. Cambridge: Cambridge University Press (pp. 67–84).

Goodwin, M. H. (1990a). *He said, she said: Talk as social organization among black children*. Blookington: Indiana University Press.

Goodwin, M. H. (1990b). Tactical use of stories: Participation frameworks within boys' and girls' disputes. *Discourse Processes* 13, 33–71.

Goodwin, M. H., and Goodwin, C. (1987). Children's arguing. In: S. U. Philps, S. Steele and C. Tanz (Eds), *Language, gender and sex in comparative perspective*. Cambridge: Cambridge University Press (pp. 200–48).

Graham, S. (2003). *Cooperation, conflict and community in computer-mediated communication*. Unpublished PhD Dissertation. Washington, DC: Georgetown University. UMI number: 3114024.

Grimshaw, A. (Ed.). (1990). *Conflict talk: Sociolinguistic investigations of arguments in conversations*. Cambridge: Cambridge University Press.

Hall, K. (1996). Cyberfeminism. In: S. Herring (Ed.), *Computer-mediated communication*. Philadelphia: Benjamins (pp. 147–71).

Hamilton, H. (1998). Reported speech and survivor identity in on-line bone-marrow transplantation narratives. In: *Journal of Sociolinguistics* 2/1, 53–67.

Harré, R., and van Langenhove, L. (1999). *Positioning theory*. Oxford: Blackwell.

Herring, S. (1992). *Gender and participation in computer-mediated linguistic discourse*. Washington, DC: ERIC Clearinghouse on Languages and Linguistics. ED345552.

Herring, S. (1994). Politeness in computer culture: Why women thank and men flame. In: *Cultural performances: Proceedings of the third Berkeley women and language conference*. Berkeley, CA: Berkeley Women and Language Group (pp. 278–94).

Herring, S. (1996a). Two variants of an electronic message schema. In: S. Herring (Ed.), *Computer-mediated communication*, pp. 81–106.

Herring, S. (Ed.). (1996b). *Computer-mediated communication: Linguistic, social, and cross-cultural perspectives*. Amsterdam: John Benjamins.

Herring, S., Johnson, D., and DiBennedetto, T. (1992). Participation in an electronic discourse in a 'feminist' field. In: K. Hall, M. Bucholtz, and B. Moonwoman (Eds), *Locating power: Proceedings of the second Berkeley women and language conference*. Berkeley, CA: Berkeley Women and Language Group (pp. 250–62).

Herring, S., Johnson, D., and DiBennedetto, T. (1995). 'This discussion is going too far!': Male resistance to female participation on the internet. In: M. Bucholtz and K. Hall (Eds), *Gender articulated: Language and the socially constructed self*. New York: Routledge.

Hymes, D. (1974). Toward ethnographies of communication. In: *Foundations in sociolinguistics: An ethnographic approach*. Philadelphia: University of Pennsylvania Press (pp. 3–66).

Kakavá, C. (2003). Discourse and conflict. In: D. Schiffrin, D. Tannen, and H. Hamilton (Eds), *The handbook of discourse analysis*. Malden, MA: Blackwell (pp. 650–70).

Lawless, E. (1993). *Holy women, wholly women: Sharing ministries of wholeness through life stories and reciprocal ethnography*. Philadelphia: University of Pennsylvania Press.

Lawless, E. (1996a). Images of God in Christian women's sermons: Finding God. In: N. Warner, J. Ahlers, L. Bilmes et al. (Eds), *Gender and belief systems: Proceedings of the fourth Berkeley women and language conference*. Berkeley, CA: Berkeley Women and Language Group (pp. 403–9).

Lawless, E. (1996b). *Women preaching revolution: Calling for connection in a disconnected time*. Philadelphia: University of Pennsylvania Press.

McWilliams, E. M. (2001). *Social and organizational frames in e-mail: A discourse analysis of e-mail sent at work*. Unpublished Master's Thesis. Washington, DC: Georgetown University.

Phillips, S. (1990). The judge as third party in American trial-court conflict talk. In: Grimshaw (Ed.), *Conflict talk* (pp. 197–209).

Tannen, D. (1982). *Spoken and written language: Exploring orality and literacy*. Norwood, NJ: Ablex.

Tannen, D. (1990). *You just don't understand*. New York: Dell Books.

Tannen, D. (1994). Gender gap in cyberspace. *Newsweek*, 16 May, 52–3.

Tannen, D. (1998). *The argument culture: Moving from debate to dialogue*. New York: Random House.

8
Language Use and Silence as Morality: Teaching and Lecturing at an Evangelical Theology College

Allyson Jule

The feminist social critic Camille Paglia (1992) discusses the power of American-style evangelical Christianity in her essay, 'The joy of Presbyterian sex,' saying there are 'Protestant looks, Protestant manners, Protestant values' central in US society today, and that being a Protestant evangelical Christian is about being in and of a specific 'tribe' with a specific, strict code of behaviour, behaviour which includes particular language habits and patterns (p. 29).[1] She goes on to suggest that all societies, including the United States, continue to need organized religions precisely because of their 'austere, enduring legacy' (p. 37); in fact, she sees it as a mistake for today's American-style evangelical Christian 'tribe' to attempt to be anything other than strict and austere because the demands of belonging and the rules of exclusion and inclusion are precisely why people, women in particular, continue to choose Protestant evangelicalism. That is, Paglia, a radical liberal feminist, believes the very austerity of religion is part of what drives many women to current expressions of evangelical Christianity. Because of the continual and rising popularity of evangelical Christianity in American life, this paper explores one specific setting within it: life at an evangelical theology college.

Laurie Goodstein (2004) of *The New York Times* reports that religion has edged its way into the forefront of American life in the last 20 years in particular. Though US history has been woven with religious issues from its inception, the interest in religion and national concerns since the 1980s has risen to now hold at 53 per cent of Americans citing religion as the key to how they vote (up from 22 per cent in 1984 – an all-time high at that point). It is now 'a normal thing' to discuss the role of religion in American society (Goodstein, p. 2). Because of Canada's proximity to the United States and the vast influence America has in the world in general, such sociological influences also impact on modern Canadian society (Stackhouse, 2002).

It is within this highly religiously charged American era that I went looking for intersections of religion, gender and language habits, specifically within the evangelical Christian community, and I am focused here on women's use of linguistic space as indicative of their role and their place in a Christian community. I locate my research in a Canadian evangelical theological graduate college because such a location allows for a discussion of religious identity and of lived practice, and of women's silence as part of currently experienced 'Protestant manners, Protestant values'.

The study

One of the most observable influences of feminism on North American Christianity in the last 30 years is the increase of women in theological education (Mutch, 2003). However, their presence in co-educational college lectures, such as the context examined here, reveals power discrepancies, even amidst these modern egalitarian times. I want to suggest that being female includes quietness as specifically demonstrative of feminine morality. Historically, and until relatively recently, theological colleges were the domain of men so that women in today's theological education have an unusual set of conditions if compared to those of the university experience in general where women's place and equality are perhaps more solidly assumed.

Little more than a century ago women were not allowed into most college classrooms. When protesting in 1910 at the admission of women to the University of Michigan, the college president said, 'We shall have a community of de-feminated women and de-masculated men. When we attempt to disturb God's order, we produce monstrosities' (in Frazier and Sadker, 1973, p. 144). Gender and religion are connected, so that much of 'God's order' is seen in the preservation of traditional masculine/feminine roles. As such, taking sociolinguist scholarship into a Christian college appeared to me an important place to explore how religion and issues of religious identity influence gendered language practices.

Women enter theological training en route to ordination – that is, en route to becoming ministers or pastors. However, many of today's American evangelicals see ordination as something still reserved for men, with women limited to supportive roles in church life (Grenz and Kjesbo, 1995). While debates within evangelical Christianity concerning the ordination of women are vigorous and dynamic (Grenz and Kjesbo, 1995, are among a host of academics writing on the subject), it is interesting to find more and more women pursuing theological graduate degrees. Regardless of the range of views on women's roles at home or in the Church, women today enroll, complete theological education, and go on to careers in many evangelical churches (Grenz and Kjesbo, 1995; Busse, 1998; Mutch, 2003; Hancock, 2003). It is also worth noting the growing feminist thought within modern evangelicalism in spite of strong anti-feminist lobby groups on the religious right, such as Focus on the Family or Concerned Women for America which promote and push 'traditional values' as central to being Christian (Coontz, 2000).

For one year, I worked on a research project at a post-graduate college. My project was to focus on the views of feminism among devout Christians living in the area. The results of the interview study are discussed elsewhere (Jule, 2004c, 2004d). However, as one trained in ethnographic methods and feminist linguistics, the year took on a slightly different focus for me, one that worked alongside the interview study. As a visiting scholar, I was able to sit in on any class of interest, either as a regular attendee or as a drop-in/on-off visitor. As such, what emerged was an ethnographic experience, one where I became a participant observer. What emerged quickly for me as a curiosity was the most used style of teaching at the theology college: lecturing, as happens in many university classrooms.

Lecturing as teaching method

Lecturing is a major part of university life. My need to appraise the method emerged from my general interest in silence in classrooms and in silence in public settings as something uniquely and most often experienced by those born female. My previous work has focused on primary classrooms and explored which speech acts teachers use to propel boys to speak more than the girls during formal-classroom language lessons (2004a, 2004b). I identify this amount of talk as 'use of linguistic space' and highlight certain classroom teaching methods as legitimating participation of the boys while serving to maintain silence among the girls.

Much research concerned with gender and its role in affecting classroom experience points to males as significant classroom participants and females as less so. Research, such as Walkerdine's (1990), Bailey's (1993), Corson's (1993), Thornborrow's (2002), and Sunderland's (1998), settles on teachers' lack of awareness of linguistic space and of how teachers themselves overtalk in the education process and, in general, give more attention to their male students (Mahony, 1985; Sadker and Sadker, 1990;

Jule, 2004a). Girls are seen as often 'passive, background observers to boys' active learning' (Spender and Sarah, 1980, p. 27). Other feminist sociolinguistic work suggests that the linguistic space used by male learners signifies and creates important social power and legitimacy (Holmes, 2001; Baxter, 1999, 2004). Who speaks tells us something about who matters inside the classroom. That men at this college participate more in question-answer time while their female classmates largely serve in the role of audience members suggests larger expectations and patterns of the community around them. That is, men contribute; women support the contributions.

Teachers and college professors talk more to their male students, beginning in the first years of schooling and on into post-graduate work. In general, female learners of all ages are rarely called upon to contribute and often find it difficult to interact with their professors. Sadker and Sadker (1990) suggest that female college students are the invisible members of the class. They suggest that one of the ways this invisibility is reinforced is through male domination of speech and through continual female silence. Kramarae and Treichter (1990) suggest that the reason women experience a 'chilly climate' in most academic settings (the college/university setting in particular) is male control of the linguistic space. Women in many college classrooms are marginalized from discourse and their silent position demonstrates and reinforces their lack of significance: they are marginal participants. That women in theology may be further silenced because of their belonging to a particular religious identity tells us something else, something more, about the relationship of religion and gender and the influence of religious views on gender performance.

Lecturing is a common teaching method at the college level. Lecturing is often used in non-university settings as well, such as public lectures held in neighbourhood libraries or art galleries. In any circumstance, lecturing is a formal method of delivering knowledge: an expert prepares the lecture well in advance, allowing for considerable research, study, and rumination as well as carefully thought through ideas and organization. People attend such public lectures for a sense of shared experience – one shared with the expert-lecturer as well as one shared with others in the audience. Lectures in such places are called 'celebrative occasions' by Goffman (1981). Frank (1995) articulates his amazement that people will disrupt their daily lives to come and hear such lectures because they have 'self-consciously defined themselves as having emotional or practical needs; they arrive already prepared to be affected in certain ways' (p. 28).

However, university lectures are part of people's daily schedules; both the lecturer and the students are usually present for obligatory reasons. The lectures are meant to disseminate knowledge for the set purposes of fulfilling the requirements of a given course. Depending on the nature of the course material, whether the course is mandatory or optional, and the size of the student group, lectures may well constitute up to 30 hours of a given course in one semester (up to three hours per week for ten weeks of an undergraduate course in most institutions). These lectures occur with such frequency that much emotional involvement is limited and are not often experienced as a 'celebrative occasion', but as a necessary practice in the university experience.

Barthes (1977) considered the university lecture in terms of politics, belonging and a location to rehearse performance discourse. While the lecturer is lecturing, the students are often silently attending to the ideas and writing notes on specific new vocabulary or content pertaining to the lesson material. The ideas expressed are in the hands of the lecturer. Much freedom is allowed concerning his or her politics, his or her power/ego issues, and his or her ability at discursive performance. As such, the lecturer has enormous control over the mood and the dynamics of the room. Lecturing as teaching method works by conveying information through summary and through elaboration – both at the discretion of the lecturer. The lecture itself is a gesture which presents the effect of universal truth. In these ways, it remains a 'celebrative occasion'. During question-answer time, students may have opportunity to publicly interact with the professor, briefly taking on the role of performer themselves by signalling investment, interest and involvement.

Goffman's (1981) ideas on the lecture differentiate between 'aloud reading', which is often perceived as more scholarly, and 'fresh talk', which is perceived as more informal though not necessarily more engaging. Barthes (1977), Goffman (1981) and Frank (1995) all recognize the lecture as a multi-layered performance. Of course, students reading the lecture material would be faster, more time-efficient, than attending class and listening to a fully performed lecture. (Perhaps listening to a cassette of the lecture while driving or cleaning the house would also be more time-efficient.) Nevertheless, the university lecture persists as a marker of scholastic participation – both attending lectures and performing lectures are part of the academic experience. Spoken delivery is also taken as candid and dynamic, more 'real', than listening to a lecture on tape or reading the notes of a lecture silently at home. Reading *A room of one's own* is one type of experience; sitting in Cambridge's Girton College's lecture hall in 1928, listening and watching Virginia

Woolf present it, would be quite another. A valuable academic lecturer is certainly one to be encountered if at all possible. As a result, the pedagogy of the lecture is 'intensely personal', even if it is personal in precisely impersonal, academic ways (Frank, 1995, p. 30).

Lecturing as power

A lecture is a mark of the lecturer's authority. What fascinates me is the way the participants themselves play the role of performers as well as the role of audience members. Their performance is briefly seen during the question-answer time of the lecture – a specific time where students pose questions to the lecturer. Lacan (1968) and his work on 'the other' as the one observed with 'the subject' as the observer influence my understanding of power relations in classrooms. His ideas propel these questions: Who is observing? Who is being observed? Which action signals and evokes power and belonging? Feminism offers various responses to these questions but it may be fair to say, in light of the vast feminist scholarship concerning pedagogy, that power largely lies in the teacher's hands. The teacher observes and the teacher performs; both signalling power. Holding the floor is the teacher's prerogative and is something which demonstrates the room's point of reference; that is, in this case, power is revealed in and created through the language practices of the lecturer. The lecturer is 'the subject' or, for Lacan, the lecturer is the 'presumed-to-know'. The lecturer is perceived as knowing and the audience members, the students, are the ones seeking the knowledge; they are the observed. What is said in lectures implicitly and explicitly hints at the lecturer's views, the lecturer's opinions on a host of issues, the lecturer's personal life and choices, including religious and moral ones. The students serve to support all of these.

Lectures in a theology college also reveal the particular context. I here suggest that lecturing, followed by question-answer periods, as is the pattern in this college, alienate the female students at this college because the feminine/masculine tendencies in classroom settings are validated and condoned by feminine/masculine patterns of behaviour within evangelical Christianity itself. With the steady increase of female theology students, it seems worthwhile to reflect on the continued high prevalence of lecturing as a common teaching method in a theology college and position it as a masculinist pedagogical tool, one that rehearses female students in feminine patterns of silence. At the same time, lecturing rehearses male theology students in masculine tendencies to dominate and control linguistic space. That is, the use of lectures in this theology college works to reinforce hegemonic masculinity (Connell, 1995) – a masculinity which insists on feminine subservience and 'reverent awe' (Gilligan, 1982/1993). Because of the transference of information/knowledge that lecturing presupposes, the silence of female students during question-answer time (a time they could speak) affirms the possibility that women behave quietly as a way of performing a specific and understood role of feminine devout behaviour: women are quiet in such a setting because their religion values their silence.

Morality as gendered

In 1982, Gilligan wrote *In a different voice* which explored various themes of gendered language patterns: a woman's place in society, gendered patterns in dealing with crisis and intimacy, and gendered patterns of expressing morality. To Gilligan, morality is closely if not entirely connected with one's sense of obligation and views of personal sacrifice. She goes on to suggest that masculine morality is concerned with the public world of social influence, while feminine morality is concerned with the private and personal world. As a result, the moral judgments and expressions of men tend to differ from those of women. In light of Gilligan's ideas, it may be reasonable to suggest that students of theology invite the suggestion that masculine behaviour is particularly connected to public displays of influence with feminine behaviour connected with more intimate, private displays. Women are rehearsed into silence for moral reasons; their silence demonstrates to others and to themselves their devoutness to God by their ability to be supportive of others. Their silence is their way of being good. Out of respect for others and for God, women are quiet. (I think this suggestion is also supported by the extensive work on silence by Jaworski, 1993.)

One would think that the current increased presence of women, the rise in feminist theology and the growth of women's ordination have significantly changed the nature of theological education. Recent research into the lives of evangelical women who chose theological education indicates that the lived experiences of these women are often painful and confusing (Gallagher, 2003; Ingersoll, 2003; Mutch, 2003). With various other religious experiences possible (including none at all), some women appear to remain in their evangelical subculture because they must experience some support and solace in their church involvement. Women who study theology say they are often dismissed as feminist for pursuing theology and are marginalized as a result. Others feel marginalized and limited and nervous about their possible future contributions; they anticipate problems though have not experienced this yet (Mutch, 2003). Canadian women in theological education report that being a woman in ministry requires 'commitment of conviction' which is carried out within a 'context of challenge' (Busse, 1998). Most cite loneliness and stress as part of their career choice and part of their theological education experiences. Nevertheless, women continue to enroll and to graduate and to go on to seek ordination in various evangelical denominations.

Much debate in theological education settles on how or if a woman can represent Christ in the church as some see the role of minister or pastor to be: to represent Christ. As such, women who choose to enter theological training at an evangelical college do so with a burden of explanation. Unlike their male classmates, they will have had to grapple with the possibility that their sex (being born female) will be a distraction at best or a continual controversy and challenge at worst.

The theology college

The particular theology college explored in this paper is located on the University of British Columbia's large campus in Vancouver, Canada.[2] The college advertises itself as 'an international graduate school of Christian studies' (school website). It also advertises itself as a 'transdenominational graduate school', not affiliated to a specific Christian denomination, though it clearly articulates evangelical Christian ideas (such as 'to live and work as servant leaders in vocations within the home, the marketplace, and the church').

Forty per cent of the student body is Canadian with an equal number (40 per cent) from the United States; the remaining 20 per cent are from other areas, including Britain and Australia as well as a few who travel from parts of Asia, Africa and Latin America. There are 350 full-time students and approximately 350 part-time students. Because the college is for graduate students, most are over the age of 25 and all have one degree behind them. Their first degree need not be theology; students come from a variety of fields, including education, medicine, law, arts, sciences. There are roughly 40 per cent female to 60 per cent male students. There are no student residences set aside for this college. Students are encouraged to find their own accommodations through the university's housing office or to live off campus. The students I spoke to all lived off campus and independently.

Students choose from a variety of Master's programmes, including a Masters of Divinity, the degree needed for ordination in most evangelical churches. There are no doctoral programmes offered, though some students continue on to pursue Doctor of Theology at larger theology colleges, such as those at Oxford, Harvard, Yale or Princeton. Most, however, enter theology college to eventually gain ordination to serve as clergy in the evangelical community. This they can do with a Master's degree.

The college employs 19 full-time faculty members: 17 are male, 2 are female. The first woman was hired in 1991; the other was hired in 2000. The imbalance of male to female faculty members, particularly in light of the male:female ratio represented in the student body, was my first clue to particular gender issues in such a setting. It is my suggestion that the religious views of this community have greatly influenced the low numbers of females in faculty positions as well as women's lack of linguistic space in the classrooms.

The classes

In light of the specifics of this community, the college is a unique location for sociolinguistic gender research. I spent eight months (one academic year) at this college as a visiting scholar. The college provided me with office space as well as access to all classes. Fifty-five classes are offered every 12 weeks (Fall term, Spring term, Summer term). The mandatory core classes are held in lecture halls which take 200 students; however, some elective courses fill at 10 or 12 and some classes run with 40–60 students.

I sat in on two of the large lectures, both consisting of approximately 200 students, both core courses. Both courses were held once a week during the Fall term. Both classes ran for three hours with one or two breaks. Both lecturers were male; the large lecture classes at the college were all taught by male faculty. The courses taught by the two women were smaller grouped classes and were not core requirements.

The male lecturers were known as senior scholars in their fields; both were well published and well known in evangelical circles. All lectures in the core courses held in the main lecture hall (such as the two in focus here) are recorded on to audiocassettes and sold in the college bookstore; hence, classroom lectures in this college serve the purpose of instruction as well as supplementary income for and promotion of the lecturer and the college.

It was clear that both lecturers, Dr Smith and Dr Jones,[3] had lectured on their material before, perhaps for years. Dr Smith was over 50 years old; Dr Jones was in his late forties. Both were of British extraction and their accents identified their ethnic background and their training. Both lectured from prepared outlines given to students. I chose these two classes because of the similarities in class size and in their use of lecture style, but mainly I chose to examine these two classes because both represented the 40:60 ratio of female students to male students as seen in the college more generally.

Dr Smith began each class with approximately five minutes of announcements, such as where to collect marked assignments, before beginning his lecture. Dr Smith had three tutorial assistants who marked weekly essay submissions. Sometimes one of the tutorial assistants (all of whom were male) would speak of these details before Dr Smith would ascend the podium. A microphone was usually clipped on to the lapel of Dr Smith's suit by a sound technician so as to record the lecture as well as to allow the entire lecture hall to hear adequately. The lecture would then begin with a two or three minute prayer by Dr Smith. Dr Smith would lecture without visible notes, though students followed along in the student packs where each lecture was provided in outline form – something purchased at the beginning of term. Most students appeared to use the lecture outlines to follow along and to write study notes throughout, filling in each section of the page. Dr Smith spoke in a clear, steady voice; he rarely used humour or personal anecdotes. At the end of the three hours (including one half-hour break), 20 minutes would be given over to questions from the students. For the 12 weeks of lectures in Dr Smith's class not one woman asked a question. Three to five male students would ask suitable questions, all higher order questions spoken into standing microphones. Each week different male students would ask questions. Their questions would last between one and five minutes. Dr Smith's responses would follow suit, generally running four to eight minutes. No female students spoke the entire term.

Dr Jones's class, however, appeared less formal than Dr Smith's. Dr Jones did not wear a suit and often arrived late and with scattered papers. Nevertheless, Dr Jones also had the aid of three tutorial assistants who often started off the class on time for him with announcements of assignments or sometimes reminders of college activities (such as the Christmas banquet ticket sales). Dr Jones usually began his lectures with a joke or humorous anecdote from his family life. Eventually an opening quick 30-second prayer was said, and Dr Jones would begin his lecture. Dr Jones used PowerPoint images which would include particular Bible passages under examination in the lecture or photographs of Biblical sites or maps. Students took copious notes; the course outline indicated the general lecture topic per week. Dr Jones gave a very long one-hour break. During the hour, Dr Jones would retreat to his office; the students ate their lunch or went to the library. Dr Jones also gave time for questions, often as long as ten minutes but averaging seven minutes most weeks. In the 12 weeks spent sitting in on Dr Jones's lectures, one woman asked a 20-second question. It was brief and answered quickly, but it stood out to me as indication of some accessibility for women in the less celebratory occasion.

In my opinion, Dr Jones was the more engaging of the two lecturers, though in my casual conversations with students, they noted very little difference. The content of each course seemed of more significance to the students than the personality of the professors, even though all students were aware of the distinguished academics in their midst and mentioned this often to me. When I asked both lecturers (casually and privately) if they noticed that only men asked questions during question time, Dr Smith said he had not noticed this; Dr Jones said he had noticed this 'years ago'. He also said, 'Women don't like to ask questions in public.'

To highlight the discrepancy of linguistic space, Dr Smith's and Dr Jones's classrooms are represented in pie-chart form in Figure 8.1.

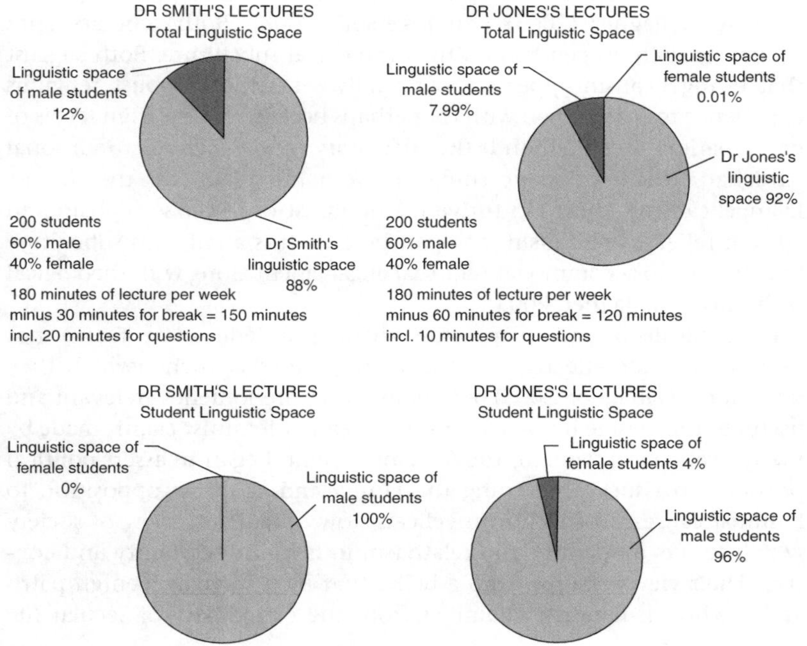

Figure 8.1 Dr Smith's and Dr Jones's classrooms

Discussion

Both Dr Smith and Dr Jones use most of the linguistic space during their lectures. This is not surprising, considering the method of instruction. However, of the remaining linguistic space, men used disproportionately more. In Dr Smith's lectures, the male students used all of the student linguistic space: 100 per cent. In Dr Jones's room, the male students used much more than 60 per cent to 40 per cent, as is the population ratio. Instead, the male students speak over 90 per cent.

When setting out the amount of talk in such form, it becomes clearer that the linguistic space, the air-time, belongs to the powerful, respected male lecturers. The male students seem to perform their masculinity by posing questions during the question-answer times allowed them. The female students are silent. Instead of talking, they appear as consisted audience members to the male performers.

Given the prominence of evangelical voices in current North American life, it is not surprising that social scientists like myself have interest in exploring the intersection of gender and language alongside religious communities. Many other researchers have undertaken numerous studies on the relationship of evangelical faith and femininity, most recently Gallagher (2003) and Ingersoll (2003). Both scholars offer robust research on gender and the evangelical subculture. Both suggest that evangelicalism appears a personally salient and robust religious experience to many, even with (or perhaps because of) the high levels of participation required and the necessary adherence to traditional Christian teachings concerning a woman's place in the home. Evangelicalism's ability to thrive in the midst of larger secularism and current religious pluralism is in part because it is a religious subculture that appears to accommodate cultural engagement along with theological orthodoxy (Gallagher, 2003).

Evangelicalism and evangelical theological education thrive not because they are effective in establishing a market niche (which they have done, Gallagher, 2003) but because they are somehow relevant and useful to the people involved. In spite of earlier feminist claims made by many evangelical women, the Christian 'right' began to assert political pressure on issues concerning the family and in direct opposition to feminist causes. In short, evangelicals now articulate a view of society which rejects modernity and relativism in favour of certainty and control. Their views emerge from a belief that men serve as 'benign patriarchs' who insulate their families from the complexity of secular life

(Gallagher, 2003). As a result, both men and women achieve morality and a resulting peace of mind by behaving in stereotypically masculine and feminine ways (men to lead, women to submit to male leadership and significance).

It may well be that women choose evangelicalism precisely as a way to find meaningful communities and to reduce the stress of navigating more complex gender roles at work and with family (Busse, 1998). Gallagher (2003), in her discussion on women in evangelicalism, suggests that evangelicals 'accommodate feminism but do so selectively' (p. 11). Gallagher suggests women remain in evangelicalism precisely because of the set roles for women. Such women find the clarity 'empowering' (p. 11). The rhetoric of a strong masculine Christianity appeals to men as well as to many women. Even organizations within evangelicalism which support and promote female ordination do so within the set dogma, offering differing interpretations of key scriptures concerning the role of women but not differing interpretations of gendered behaviour; men are to be strong and rational, women are to support male 'headship' even if they are ordained.

These complexities within evangelical circles, specifically that one could be a female ordained minister and still remain a woman committed to submission as a key moral and gendered behaviour, suggest that women in this college must manage the contradiction within these ideas. These women have proceeded to pursue theological education, not for reasons of liberation or female emancipation from male domination in the Church, but as a way to serve the Church with their gifts of service. Though some women may have difficulty in such a context, many appear to remain and further invest themselves precisely because of a sense of calling. They remain in their 'context of challenge' because of their 'commitment of conviction' (Busse, 1998). They work out their gender roles within a larger framework of male leadership and domination. Even if ordained, women see their roles as supportive and not leadership-driven.

That lecturing is used in such old-fashioned ways within this theology college (and my guess is in many others as well) suggests a clash of conservative/spiritual masculinity with pro-feminist/social justice masculinity within university education (Clatterbaugh, 1990; Skelton, 2001). Such a domination of linguistic space is what Skelton calls 'the school and machismo': that the ways males experience or exploit educational opportunities are 'skills' which males in society 'learn to develop' (p. 93). Men are rehearsed into speaking roles; women are rehearsed into listening roles.

Conclusion

I agree with Paglia (1992) in that there are distinctive 'Protestant looks, Protestant manners, Protestant values' and that these codes of belonging to Protestant evangelical Christianity are attractive and seductive to many men and women. That women choose to belong and to support a religion which views stereotypically gendered roles as desirable might explain their alarming silence throughout the courses I observed at this evangelical theology college. The women remained silent all term. I believe that such specific manners and values are part of their being seen as devout in this community.

Women's roles are supportive roles, even if appearing to reach for the top levels of church governance by enrolling in Masters of Divinity programmes. The 'Protestant looks, manners, and values' seem to include feminine silence. Though the evangelical world has competing debates within it concerning the role of women in the home, in society and in the Church, it appears to be the case at this college that female presence has not meant an upset in 'God's order' and that women serve as supportive listeners to the larger male-dominated linguistic space of the lectures.

If a century ago there had been a fear that women in theological education would 'de-feminate women and de-masculate men', this fear appears an unnecessary worry because, even when present, women continue to behave in quiet, submissive, supportive ways in these college classrooms. A masculine style of seeking public influence and participation as a way to be moral seems also at work at this college; the men acted the part of knowing, belonging to and participating in power, while the women served the part of audience. These patterns are so commonly seen in other pedagogical research in classrooms that the findings are not surprising. What this research does point to is cultural and historical threads which have appeared as patriarchal but which may reveal women consciously or subconsciously colluding in such patterns. Many women remain in evangelical Christianity; women participate in theological education; lecturing as pedagogical tool is a popular teaching method at this college; and women continue to serve well as audience members in their own educational experiences. These are the Protestant manners and values at work in American society. The lecturer as performer is well-received and well-supported by college life in general. The male privileging of this style is one not well-interrupted and my guess now would be that both the men and women who belong to specific religious groups would defend gendered behaviour as 'God's order'. The popularity of evangelicalism, particularly in American life, gives rise to feminist scholarship in a search for understanding women's experiences in public life.

Notes

1 The term 'evangelical' is used to refer to those of Protestant faith who are Pentecostal, fundamentalist or mainline liberal – terms articulated by Gallagher, 2003. She also suggests evangelicals are generally anti-feminist and anti-big government; they hold these views because of their perceptions of what 'the Bible says' and they promote the 'Good News' to convince others of their views.
2 There are several theology colleges on the University of British Columbia's campus. The one examined here will remain nameless for reasons of anonymity.
3 The names are fabricated to protect anonymity. Also, 'Doctor' is usually the title used for professors in Canada, indicating a PhD as well as professor status. To be called 'Professor' may indicate no PhD and, hence, less credentials.

References

Bailey, K. (1993). *The girls are the ones with the pointy nails*. London and Canada: Athouse Press.

Barthes, R. (1977). Writers, intellectuals, teachers. In: *Image-Music-Text*, trans. S. Heath. New York: Hill (pp. 190–215).

Baxter, J. (2004). *Positioning gender in discourse: A feminist methodology*. Basingstoke: Palgrave Macmillan.

Busse, C. (1998). *Evangelical women in the 1990s: Examining internal dynamics*. MA thesis. Briercrest Bible Seminary, Caronport, Saskatchewan.

Clatterbaugh, K. (1990). *Contemporary perspectives on masculinity: Men, women and politics in modern society*. Washington, DC: Westview Press.

Coates, J. (1993). *Women, men and language*. 2nd edn, New York: Longman.

Coates, J. (2003). *Men talk*. Oxford: Blackwell.

Connell, R. (1995). *Masculinities*. Cambridge: Polity Press.

Coontz, S. (2000). *The way we never were: American families and the nostalgia trap*. New York: Basic Books.

Corson, D. (1993). *Language, minority education and gender*. Clevedon: Multicultural Education.

Daly, M. (1968). *The church and the second sex*. New York: Harper & Row.

Doriani, D. (2003). *Women and ministry*. Wheaton, IL: Crossway Books.

Eckert, P., and McConnell-Ginet, S. (2003). *Language and gender*. Cambridge: Cambridge University Press.

Edelsky, C. (1981). Who's got the floor? *Language in society* 10 (3), 383–422.

Frank, A. W. (1995). Lecturing and transference: The undercover work of pedagogy. In: J. Gallop (Ed.), *Pedagogy: The question of impersonation*. Bloomington: Indiana University Press.

Frazier, N., and M. Sadker (1973). *Sexism in school and society*. New York: Harper.

Gal, S. (1991). Between speech and silence: The problematics of research on language and gender. *Papers in pragmatics* 3 (1), 1–38.

Gal, S. (1995). Language, gender, and power: An anthropological review. In: K. Hall and M. Bucholtz (Eds), *Gender articulated*. New York: Routledge (pp. 169–82).

Gallagher, S. K. (2003). *Evangelical identity and gendered family life*. London: Rutgers University Press.

Gilligan, C. (1982/1993). *In a different voice*. Cambridge, MA: Harvard University Press.

Goffman, E. (1981). The lecture. In: *Forms of talk*. Philadelphia: University of Pennsylvania Press (pp. 160–96).
Goodstein, L. (4 July 2004). Politicians talk more about religion, and people expect them to. *The New York Times*. Weekend (p. 2).
Grenz, S., and D. M. Kjesbo (1995). *Women in the church*. Downers Grove, Ill: Intervarsity Press.
Griffith, R. M. (1997). *God's daughters: Evangelical women and the power of submission*. Berkeley: University of California Press.
Hammersley, M. (1990). *Classroom ethnography*. Buckingham: Open University Press.
Hancock, M. (Ed.) (2003). *Christian perspectives on gender, sexuality, and community*. Vancouver: Regent College Publishing.
Holmes, J. (2001). *An introduction to sociolinguistics: Learning about language*. New York: Longman.
Ingersoll, J. (2003). *Evangelical Christian women: War stories in the gender battles*. New York: New York University Press.
Jaworski, A. (1993). *The power of silence*. London: Sage.
Jule, A. (2004a). *Gender, participation and silence in the language classroom: Sh-shushing the girls*. Basingstoke: Palgrave Macmillan.
Jule, A. (2004b). Speaking in silence: A case study of a Canadian Punjabi girl. In: B. Norton and A. Pavlenko (Eds), *Gender and English language learners*. Virginia: TESOL Press (pp. 69–78).
Jule, A. (2004c). Gender and religion: Christian feminism. Paper presented at the Religion and Society Conference, Point Loma University, San Diego, California (23–26 March).
Jule, A. (2004d). God's daughters? Evangelical women speak for themselves on feminism. Poster presented at IGALA3, Cornell University, Ithaca, NY (5–7 June).
Kramarae, C. (1980). Gender: How she speaks. In: E. B. Ryan and H. Giles (Eds), *Attitudes towards language variation*. London: Edward Arnold (pp. 84–98).
Kramarae, C., and P. Treichler (1990). Power relationships in the classroom. In: S. Gabriel and I. Smithson (Eds), *Gender in the classroom: Power and pedagogy*. Chicago: University of Illinois Press (pp. 41–59).
Lacan, J. (1968). *The language of the self: The function of language in psychoanalysis*, trans. A. Wilden. Baltimore, MD: Johns Hopkins University Press.
Mahony, P. (1985). *School for the boys: Co-education reassessed*. London: Hutchinson.
Matthews, W. (1991). *World religions*. St. Paul, MN: West Publishing Company.
Mutch, B. H. (2003). Women in the church: A North American perspective. In: M. Hancock (Ed.), *Christian perspectives on gender, sexuality, and community*. Vancouver: Regent College Press (pp. 181–93).
Paglia, C. (1992). The joy of presbyterian sex. In: *Sex, art, and American culture: Essays*. New York: Vintage Books (pp. 26–37).
Porter, F. (2002). *Changing women, changing worlds: Evangelical women in church, community and politics*. Belfast: The Blackstaff Press.
Ruether, R. R. (1998). *Introducing redemption in Christian feminism*. Sheffield: Sheffield Academic Press.
Sadker, M., and D. Sadker (1990). Confronting sexism in the college classroom. In: S. Gabriel and I. Smithson (Eds), *Gender in the classroom: Power and pedagogy*. Chicago: University of Illinois Press (pp. 176–87).

Scanzoni, L. D. (1966). Women's place: Silence or service? *Eternity* 17 (February) 14–16.

Skelton, C. (2001). *Schooling the boys: Masculinities and primary education*. Buckingham: Open University Press.

Spender, D., and E. Sarah (Eds) (1980). *Learning to lose: Sexism and education*. London: The Women's Press.

Stackhouse, J. G. (2002). *Evangelical landscapes*. Grand Rapids, MI: Baker Book House Company.

Storkey, E. (2001). *Origins of difference: The gender debate revisited*. Grand Rapids, MI: Baker Book House Company.

Sunderland, J. (1998). Girls being quiet: A problem for foreign language classrooms. *Language Teaching Research*, 2.

Thornborrow, J. (2002). *Power talk*. London: Longman.

Walkerdine, V. (1990). *Schoolgirl fictions*. London: Verso.

9
The Children of God Who Wouldn't, but Had To*

Annabelle Mooney

Writing about groups that are typically branded as cults is like walking a high wire; except everyone wants you to fall. The group you choose to examine will certainly never consider itself a 'cult' (neither should it). Those who are interested non-members (often ex-members) see the group as exactly a 'cult'. In this chapter, I will not be engaging with the question of what constitutes a 'cult'. It seems to me, that according to the usage to which the term is put, no group is actually a 'cult'. I am not, however, going to suggest that groups, especially perhaps religions, are always positive experiences for members. It is clear that they are not.[1]

This chapter is in one way a historical account. The Children of God no longer exist as such, having been re-formed and renamed The Family (sometimes also known as The Family of Love). The question of whether this has been a change in name only is not one I will address at length here.[2] The text under examination in this chapter, however, represents women (or more correctly, the gendered female) as powerful and worthy only in their giving up of self and power. They are only valid subjects if they transform agape into eros; in turn, they need to transform themselves such that this is not a contradiction. That is to say, they are recognized as subjects only when they sublimate agape into eros and love into obedience.

It is difficult to know how life in The Family differs from that in The Children of God. What is clear in this text is that part of agape is eros. That is, the selflessness which is part of the love of agape extends to being

* I would like to gratefully acknowledge and thank the help of Mercy, an ex-member, who guided me through doctrinal and historical issues, and answered many questions. Her story can be read at www.exfamily.org. While this account underpins some arguments here, specific help is referenced with 'personal communication' which took place July 2004 via e-mail.

selfless when it comes to acts normally associated with eros. The self too is discounted in favour of this selfless giving of emotional and sexual favours. Salient in terms of gender is that it is the feminine and feminized member who submits. Thus gender and power coalesce in the sexual frame.

Because of the witnessing (and arguably exchange) purposes to which sex has been put in the group, I argue that it is possible to see the female as commodity (in line with Irigaray, 1985). At the same time, the feminized quality of sacrifice is both valorized and sexualized. It hardly needs stating that it is sexualized for men (rather than women) and in particular for the male leader (who is essentially divine). It seems to me that the illustrations of women, in particular, are pornographic.

While the material here looks rather extreme, especially for a group that sees itself as Christian, the logic of the position is not difficult to understand. It is worth stressing the importance of understanding groups that seem beyond the pale. In fact, it seems to me that a great deal of harm is done (to women especially) if and when they choose to leave a group which is dismissed as 'cultish' in main stream society (see Boeri Williams, 2002).

It seems to me that it is also important to point out that this commodification of the female body is not something that is limited to The Children of God. Sexual and gender politics in mainstream society are not radically different from the kinds of views put here. That is not to say that these views should be endorsed, rather that, upon reflection, they seem rather less radical than one might at first think. However, it is well to bear in mind that the dynamics of a closed society (such as this group) create different pressures on individuals; especially when the leader is essentially divine.

History

Wangerin sees the Children of God as a 'symbolic rebellion against American capitalist culture' (1993, p. 1). It was founded in 1968 by David Berg, later called Moses or Mo or even 'dad'; though by the second generation (offspring of original members) he is referred to as 'Grandpa'.[3] Originally, Berg was merely an evangelical preacher, though it was not long before he proclaimed himself a prophet and the group stood apart from any traditional religion. After his death, his mistress Maria took on the leadership of the church.[4] The Family presents itself as a group distinct, though directly descended, from the Children of God. However, ex-members maintain that the group is the same. Indeed, in Mo Letter 663, 'Happy Birthday! – RNR[5] Rules – A Compilation' in paragraph 3,

Berg writes, 'Try to get away from being called the "Children of God" where the name is not helpful nor legally necessary. Simply call yourselves the new Family of Love' (1978b).

The Children of God were controversial because of their attitudes to sex. Sexual enjoyment was seen as a part of life.[6] They believed, for example, that Jesus had sexual relations with Mary and Martha.[7] The year 1976 saw the introduction of the doctrine of flirty fishing (FF),[8] though it was discontinued in 1987 because of adverse public reactions and the sexual health of the community.[9] It involved young women of the movement using sex as a form of religious outreach based on Matthew 4:19, 'Follow me, and I will make you fishers of men.'[10] Because this flirting was to bring people to God, it was considered acceptable. Sexuality is seen as a gift from God and thus something which can be legitimately used to secure members. It should be noted that 'Though men were not as effective in FFing, they did participate and occasionally were able to bring in new female disciples' (Chancellor, 2000, p. 16). Williams (1998) also points out that while 'Men were legitimately allowed to sexually recruit women into the family, and although this happened sometimes, it seemed to be much more time-consuming, and became a rare event' (p. 133). Mercy concurs: 'Although it [doctrine of FFing] was directed to the women and it was the women who went to the bars dressed seductively, men could also do it. However, it was really the women who were to bear the brunt of this new "ministry" and the women who eventually were to bring in quite a bit of money that propelled the work into a financial upswing during the heydays of the late 1970s and 1980s until a member died of AIDS and the practice was sharply curtailed' (2000, on-line).

The text which I will be examining is one of a much larger collection of texts called the 'Mo Letters'. There are in excess of 2500 of these letters.[11] They are sent out to all branches of the movement and were standard reading material for all members (Enroth, 1977, p. 43). In this they serve to introduce members to Mo and also to give the group an individual corpus of readings. They thus provide a unique text for the movement and because of this a way for members from different communities to speak to each other. These letters are the way in which the particular beliefs of the movement are articulated and disseminated. The letters are also available to the general public.[12]

Berg died in November 1995. At present, there are about 14,000 members, a third over 21 years. Bozeman reports that morale appears to be high in the group and 'The Family appears to be well prepared to continue propagating its distinctive doctrinal message and lifestyle for the foreseeable future' (1998, p. 129).

The text under examination here is a Mo letter from 1978 called 'The Girl Who Wouldn't.' The letter was chosen because of its explicit discussion of sex with respect to a female member who wouldn't have sex with another (senior) female member. The way in which the girl who wouldn't is spoken to and about suggests that what is at issue here is not so much gender, but power; or rather the latter determines the former.

The letter: the girl who wouldn't

This letter was sourced from an ex-member organization (see note 11). Apart from two versions of the letter (one 'full' version and one condensed), the web site also provides a commentary. In this analysis I refer to both the full version (FV) and the short version of the letter (SV). The full version appears to have been circulated at the time (i.e., 1978) in direct response to letters written to Berg by a senior member of the group, Lori, and her assistant, Toni. These letters, with comments inserted by Berg (to be discussed presently) are published at the start of the full version. This version also includes an illustration. It is not clear whether the letters from Toni and Lori were included in the condensed version; though given that the shorter version is in many ways more general, with no specific mention of the contents or authors of the original letters to Berg, it seems safe to assume that they were not.

The long version is more than four times longer than the short. It includes the initial letters and 100 paragraphs of responding text. The short version is only 25 paragraphs long and does not refer to Lori and Toni in the same way as the long version.

The letters from the girls

Lori reports that working with Toni has been good and 'our communications have been excellent ... THE ONLY THING THAT HAS MADE IT KIND OF DIFFICULT IS THAT WE NEVER GOT TOGETHER PHYSICALLY.' Toni's letter laments that she has 'failed Lori in a lot of ways by not being able to be her mate'. Toni writes that she can 'burn free' with family work and FFing (flirty fishing) but lacks the faith for being physical with Lori, even though she sees nothing wrong with it. It is significant that both letters are presented as being written in the style of Berg's own Mo letters. They all, for example, use capitals, presumably for emphasis.

The response from Berg, with some interjections from Maria, makes it abundantly clear (even in the short version of the letter) that this refusal is not appropriate. The terms in which this disapproval is phrased makes it clear that Toni is disobeying the word of God and the word of Berg.

Toni's refusal to have sex with Lori is also represented as symptomatic of selfishness of a more general kind which is not acceptable within the movement. While this is certainly a gendered issue, what seems to be at stake here is power and a right to sex. Wanting sex is not selfish; withholding it is. The former is gendered male; the latter female.

In this first part, the letter itself is under consideration. I deal first with some aspects of orality present in the letter, including interjections from Maria. The layout of the text and the attached illustration are then briefly considered. The central argument of the letter relies on a minimizing of the sacrifice that Toni should have made, and a maximizing of the error that was committed in not making this sacrifice. This is cast in a familiar 'all or nothing' mode. A section focusing more on gender follows.

Orality and conversation

As noted, Mo letters are staple texts of the movement. It is not necessary to look beyond this letter to see that this is the case. In several places Berg exclaims that Toni hasn't read 'the Letters' and isn't following them. These letters are Mo letters and appear to trump the other text of the movement, the Bible. By this I mean that the Mo letters offer the official interpretation of the Bible. It is an interpretation which is not open to question by those in the group. Thus while the Bible may not be transparent to all, it is transparent to Berg. It is worth noting the capitalization of 'Letters', which is conventionally aligned with the capitalization of the 'Bible'.

Even in a written form the letters manage to capture something of a physical voice. This letter is no exception and uses the techniques commonly deployed by Berg. In this respect, it is interesting to note that the letters were often read out loud to communities, adding another level of 'vocal authenticity' to the letters. Two strictly visual cues signal vocality; exclamation marks and capitals. In the long version of the letter, which is 100 paragraphs long, there are 235 exclamation marks. In the short version (which is only 25 paragraphs long) there are 73 exclamation marks. Exclamation marks are common in Mo letters and part of Berg's written style. While their widespread use starts to look hysterical, this seems to be exactly what Berg is after; 'I'M A HUNDRED-PERCENTER! I CAN'T STAND COMPROMISERS! I CAN'T STAND HALF HEARTED PEOPLE!' (para. 17 LV; para. 4 SV). Certainly there is nothing half-hearted about Berg's typographical conventions here. Although his use of capitals is less pronounced in the condensed version of the letter. In fact, in the letters posted on the official website, this typographical convention has been removed.

In the long version, the first sentence or clause of every paragraph is capitalized. In the shorter version, capitals appear to be used only for vocal emphasis; roughly once every five paragraphs. Berg's lexical and syntactic choices also indicate a more oral style; repetition, rhythmic structures, in-group and non-standard vocabulary (i.e., lexemes usually used for speech are written, for example 'cuz' for 'because') as well as sentence fragments all contribute to this.

This letter, however, also has some conversational structure in a number of ways. First, Berg's letter is a response to those of Toni and Lori. However, in a later (though related) letter, 'Women in Love', Berg makes clear that he will not continue to be a problem solver for the entire membership (para. 22).

The second way in which the text is conversational (though not on an equal and collaborative footing) relies on Berg's editing of Toni and Lori's letters. These are reproduced at the start of the long version, and Berg has inserted comments flagged with the tag 'Dad'. Toni writes:

> I'VE GONE THOUGH IT SO MUCH THINKING HOW SELFISH AND WITHHOLDING I MUST BE and maybe I am. Please tell me if you think I am. ***(Dad: Amen! – You are!)***. But most of all I just want to be obedient to God's will ***(Dad: So why weren't you when it was needed?)***

These insertions forestall any sympathy with Toni. Even before one can read her letter in full, Berg's voice has been inserted in bold italics. This signals very clearly whose voice has priority. Further, the power that he has to interject, and thus interrupt another's text, is indicative of his leadership power. Berg, as the 'father' of the group ('Dad'), is at liberty to reconstruct the letters in such a way as to control their message.

The third sense in which this is a conversation is the co-present vocality of Maria (Berg's mistress and now leader). This conversational structure is more a co-production of text, and may be understood as code switching, in the main part of the letter. In the letter itself, Berg speaks for himself and Maria, explicitly including her in his opinions; 'as far as Maria and I are concerned' (para. 6, LV) and so on. However, Maria sometimes also offers her own voice, but none the less she only makes five short contributions in the long letter, usually supplying Biblical support. Maria argues, for example, that had Toni started to be physical with Lori, the desire would have come:

> (MARIA: AND I THINK THE LORD WOULD HAVE GIVEN HER THE LOVE THEN, that's what usually happens. 'As they went they were healed'. – Lk. 17: 14.)
>
> (SV 7; LV 24)

Maria's use of 'usually' suggests that she has some experience of this. In terms of gender roles, however, Maria is clearly subordinate to Berg. The fact that she has very little to say, that she is included in Berg's words and (perhaps most of all) that her contributions are enclosed in parenthesis, demonstrates this clearly. Further, Maria usually contributes words of the Bible. She is not speaking as such, she is speaking for the authority which gives Berg his power.

Nevertheless, involving Maria in this exchange has the benefit of making the argument not look like a gendered one. In terms of gender and power, however, Maria as partner of the leader is part of the 'royal family' of the movement.[13]

Indeed, Berg himself asserts that he has had to submit. In writing this he attempts to occupy the female role. He writes:

> 9. A LOT OF THE TIME I DON'T EVEN WANT TO or like it or enjoy it because I'm tired, and had enough, etc. Sometimes it's even with people whom I'm not even naturally very fond of or that I'm not very crazy about their personality or their type or their shape or their manner etc.

It is difficult to believe that as leader of the movement this is the case.

Format and illustrations

As noted, the arrangement of this Mo letter is similar to others. Each paragraph is numbered sequentially. This form appears to capitalize on associations with Biblical presentation and makes it possible to cite portions of the letters, in the same way that Bible passages are quoted; each letter also has a unique number code as well as a title. However Wangerin also notes that it, and the general style of the letters, made 'reading them, or following along while others read them aloud, easier' (1993, p. 128).

The most striking thing about this Mo letter is the inclusion of an illustration at the top of it. A line drawing with the capitalized title 'THE GIRL WHO WOULDN'T!', it shows two female figures in a bedroom. A long-haired woman is standing in the foreground with her back to her companion. She is wearing a short nightdress, has her arms folded across her chest and her eyes shut; her chin is raised in a pose of defiance. A short-haired woman is naked, in the background, on a bed sitting with her back against the headboard and her arms stretched out to her sides. One of her knees is bent.

It is clear that the woman in the foreground is the girl who wouldn't. It is also pretty clear from the picture (not least the clothing and lack thereof) what she wouldn't do. While the picture quality (in the version I saw) is not excellent, there is a case to be made for the woman in the background being posed in a typical crucifixion position. Certainly the raising of one knee means that the illustration is not x-rated, but it is also a common representation of Christ's position on the cross. Additionally, with her arms stretched out to the side, the overall effect is even more typical. This reading is given further credence by the way in which Berg links the 'Lord's worker' (Lori) to the Lord. 'Don't tell me you're good for His work if you're not good for his workers' (LV 36); and 'When she refused you, she refused the Lord' (SV 21). Thus essentially, Toni is turning her back on the Lord. The nakedness of Lori is at once sexual and vulnerable; again it evokes Christ's nakedness on the cross. But because of this link with the Lord, it is also a position of power.

In terms of the representation of women in the movement generally, cartoons like this are extremely telling. While I have by no means seen a complete set of Mo letters, of all that I have seen the representations of men and women differ from each other, but are consistent in themselves. In short, the men are nearly always clothed (and usually bearded), while the women are always naked or in a state of undress (often with their clothes ripped and hanging off them).[14]

Bainbridge writes that 'The original Mo Letters and many of the later publications carry lively illustrations like those found in comic books' (2002, p. 75). Certainly the pictures are in the style of comic-book illustration, being line drawn and so forth. But to call them 'lively' when many are erotic representations of women dismisses this influential mode of communication. This sustained representation of women is, it seems to me, particularly significant and troubling. Not only are the women usually represented in a way which reveals their bodies, they are sometimes in situations of pain. In collections I have personally seen 'cartoons' of women crucified, their wounds dripping with blood, or sitting naked on a large fish hook (as a representation of flirty fishing). 'Heaven's Girl', a cartoon series directed at young people, was particularly disturbing (though now withdrawn from circulation). If men are naked, they will be embracing a woman. It seems to me that these illustrations are pornographic. I will return to this at the end of the chapter. Further, while men could FF, they were not asked to. Rather, they were asked to be God's pimps.[15]

While the sexual doctrines allegedly apply equally to the sexes, these illustrations suggest that this is not the case. Further, ex-member accounts also suggest that it was women who submit and men who desire. It is a heterosexual dynamic that predominates (not withstanding this letter). In fact, a later letter deals specifically with lesbians. Called 'Women in Love' it states that while there is nothing 'wrong' (para. 8) with lesbianism, and while it's not 'EXPLICITLY FORBIDDEN, SUCH AS IT IS WITH MEN AND MEN' (para. 2), it is 'subnormal' at least in as far as it can't 'bear the physical fruit of children' (para. 8). (This kind of argument was used against Lori (by then known as Keda) in a letter called 'Keda's Problem'.)

The letter starts by saying that there doesn't seem to be anything wrong with women having sex together. And yet towards the end, Berg writes, 'ANY GIRL WHO DOESN'T LIKE RELATIONSHIPS WITH MEN AT ALL AND ONLY LIKES GIRLS HAS SOME KIND OF A PERVERTED SPIRIT!' (para. 100). Such women are apparently 'BORN WITH EXCESSIVE MALE HORMONES AND CHARACTERISTICS, and are almost more male than female, and therefore have an unusual desire for other women rather than men' (para. 101).

Indeed, the imperative to have (many) children can be seen as subjugation of women. It is a society such that 'Woman lives her own desire only as the expectation that she may at last come to possess an equivalent of the male organ' (Irigaray, 1985, p. 24). Further, the power of the male can be understood as underpinning the forbiddance of male homosexual activity. 'Once the penis itself becomes merely a means to pleasure, pleasure among men, *the phallus loses its power*' (Irigaray, 1985, p. 193).

In 'The Girl Who Wouldn't' it is women (in particular and in general), not men, who are chastised for withholding sex.

> IT REMINDS ME OF SOME OF OUR SELFISH SISTERS IN THE FAMILY WHO HAVE GONE ALL THESE YEARS WITHOUT HELPING ONE SINGLE BROTHER, not giving one of them a tumble, not once!
>
> (para. 25, LV)

In terms of 'The Girl Who Wouldn't' there is another point made in 'Women in Love', right at the beginning of the letter, which should be noted. Berg writes that with lesbian relationships 'All the same rules apply' (para. 3). That is:

> It must be **real love**, not just a sexual lust. It **must** be **love**. You **must** do it in **love**. It must be done with **understanding**. It must be done by **faith**, not under condemnation, knowing your liberties in the Lord. And it must be done with mutual **consent**.
>
> (para. 3)

The 'must' may look like a requirement, but in fact it turns out to be a command. The consent is not consent to the particular situation, but to the authority of Berg to make commands.

Central argument

The main argument of the piece should already be clear. Toni should have sacrificed herself to Lori's sexual needs. The use of orality already discussed, as well as the priming features of the illustration and Berg's comments inserted in the letters from Lori and Toni make it clear from the start where Berg stands. He is emphatic about his views in the letter, however. This is especially important for the shorter version which lacks some of these priming features.

Essentially, the argument works by minimizing the sacrifice Toni has to make and maximizing the cost of not making the sacrifice. This devolves into a familiar all-or-nothing script which is closely related to an us–them othering. In this case, the other is explicitly the Devil (with whom Toni is aligned). The argument also works by turning the particular to the general. The minimization of self is part of the minimization of sacrifice and is clearly a strategy of control. Further, sexual desires are treated as though they are requirements. Apart from a virtual collapse of agape into eros, it is also clear that self can only be a source of action in a limited range of spheres. Self can initiate sex, for example, but it cannot refuse it.

Minimum and maximum

This letter is one which chastises. For the minimization of the sacrifice to work in such a case, much has to be made of the bad consequences of not having made the sacrifice first. Thus in the short version of the letter, it isn't until paragraph 17 (of only 25) that Berg makes small of the action Toni had to commit (in the long version it is at paragraph 61 of 100):

> Just because she didn't want to do one little thing, make one little sacrifice, one little moment of physical self-denial – surrendering her mere flesh for a moment.

Not only is the action itself a 'little thing', her body itself is 'mere flesh'. Put this way, the act of sex starts to look like merely a physical act of strength or endurance. In fact, in paragraph 6 (SV; 23 in LV) Berg writes:

> Even if it rubbed her the wrong way and she didn't like it at all, she should have gritted her teeth and borne it, just like any husband or

> wife or any mate or any FFer has to do sometimes, even if she hated it!

It's worth bearing in mind that Toni is an FFer and apparently has no trouble surrendering herself to men for the purposes of outreach. We can understand this as a distinction (for Toni perhaps, but not apparently for Berg) between active and passive sexual sacrifice. It's difficult to see how in a lesbian sexual situation she could have 'gritted her teeth and borne it'. Indeed, Berg questions Toni's commitment to flirty fishing, remarking, 'Then I heard she's been an FFer. I wonder how far her FFing went – if it went to the bed? I don't know' (LV 37: not in SV).

While the sacrifice is little, the cost of **not** making the sacrifice is extreme. Berg is explicit about this. In the long version he is rather voluble. In paragraph 22 of the short version he writes (81 in LV).

> In this one 'little' thing she failed God and you and us and the Family! But she never realized how important it was and what a big thing it was, that God was testing her to see if she was willing to give her all, and she wasn't – so she failed!

In many ways here the point is sound. If we take sex out of the argument, many Christians would agree that to fail to do a little thing is to fail God. Without paying a great deal of attention to this, it is worth noting the absence of active forgiveness. It is possible for the sinner to 'repent and stop' (para. 24 SV; 93 in LV), but there is no active offering of grace.

All or nothing/us them

Not only is there no active offering of grace, there is a complete banishing from the group and from God. Berg says again and again in the long version that he wouldn't care to have Toni on his team. He compares her to the Devil, calls her a 'self-worshipper, a pride worshipper' (para. 28 LV; para. 9 SV), 'selfish and independent and unsacrificial and disobedient and rebellious and stubborn' (para. 30 LV; para. 11 SV). It is clear, then, that not to sleep with Lori is enough for all Toni's good works to be worth nothing. Berg argues that Toni can't be good at caring for children if she can't sacrifice herself. In short, 'there's no excuse for not having faith for anything!' (para. 7 SV). Not being able to cope with the demands of the movement is lexicalized as 'stumbling' and is due to personal weakness.

Particular to general

In the short version especially, the particular case of Toni and Lori has been translated into a more general case. Thus it can be read as a more general warning that members should not refuse sexual advances, as a request from a member is a request from God. Further, the letter can be seen as more generally about following orders and following the Letters in particular.

The second paragraph of the short version (para. 8 in LV) makes this abundantly clear. Berg says that his mother used to say:

> Don't tell me you believe in something if you're not doing it! You believe in prayer as much as you pray and you believe in witnessing as much as you witness!

Between the first and the second version, Toni becomes not just a girl who wouldn't sleep with a woman, but a girl who wouldn't do anything as set out in the Letters.

In the long version, the move from particular to general is similar. It is clear that this is not just a chastising of Toni, but of anyone (women in particular) who chooses not to sacrifice self. The public humiliation of Toni doesn't stop at her individually; it includes, for example, all Scorpios:

> 39. SHE SOUNDS LIKE A TYPICAL SELFISH, SELF-CENTERED SCORPIO! Many of the Scorpios I know are about as selfish and as independent as they come! They don't even like to get married! They don't like to have any obligations to anybody but themselves.
>
> (LV)

It is not only Toni who is taken to task, but any members who are selfish and disobedient (Scorpios in particular beware).

Minimization of self and eros/agape

The obedience described in this letter is absolute. 'Helping out' other members sexually is seen as a perfectly reasonable sacrifice to make (para. 21 LV). There is no room for independence or freedom of choice. Berg again makes this explicit:

> AS FAR AS I'M CONCERNED SHE DOESN'T BELIEVE WHAT I WRITE BECAUSE SHE DOESN'T OBEY WHAT I WRITE! She has withheld herself selfishly and independently from you.
>
> (para. 22 LV)

More interestingly, perhaps, this minimization of self appears to involve a conflation of eros and agape. That is, if one loves another human, then one should be sexually available to them. In the doctrine of Berg, as every person is a neighbour, then everyone should be sexually available. While in 'Women in Love' a careful distinction is drawn between types of love, the realization that one usually leads to the other is admitted:

> 35. **'PHILEO' OR BROTHERLY (OR SISTERLY) LOVE DOES NOT HAVE TO ALWAYS WIND UP IN SEX! Brotherly** love, **Godly** love – these do not have to necessarily always wind up in each other's arms sexually. They're inclined to, but let's face it, if God's forbidden that between **men**, then they better lay off! But, if He **hasn't** forbidden it between **women**, I don't see why not.

Brotherly/Godly love thus conflates phileo with agape; and this is inclined towards eros. In 'The Girl Who Wouldn't', the separation of the sexual from the emotional (the eros from the agape) is illogical from Berg's point of view. As mentioned, every person should be treated as though they are the Divine (God or Berg):

> IF SHE WOULDN'T YIELD TO YOU WHEN YOU'RE REPRESENTING US, HOW COULD SHE YIELD TO US PERSONALLY? How can she say, 'I love and respect Lori with all (?) my heart' when she's unwilling to give you her body too! Don't tell me you can give your heart without giving your body! Ridiculous!
>
> (para. 58 LV)

Gender

Gender relations in this text can be understood as power relations. Even without considering the 'Women in Love' letter, it's easy to see that Lori is given a masculine role. 'As soon as she has any relationship with another woman, she is homosexual, and therefore masculine' (Irigaray, 1985, p. 194). Lori is more senior, she desires sex, and is aligned semiotically with Christ. Toni, on the other hand, is not only female, she is a 'girl'.

In a sense, all members are gendered female as they are all brides of the Lord. The 'Loving Jesus Revolution', a move begun in the mid-1990s is explicit about joining sex and worship together. Amsterdam remarks:

> Now, it is pretty easy for the women to say 'You are my lover' to the Lord. But the men, well we are quite opposed to homosexuality.

> The Lord told us in a prophecy that it is no big deal. He calls everyone His bride. It is not a *visible* thing, there is no male or female in Christ Jesus. It is like role playing. The men will play the role of the bride.
> (Chancellor, 2000, p. 147)

Whenever sex is involved, gender is an issue. The practice of flirty fishing is a good example of how women are sexualized for the work of the movement and men are not. Men are consumers and women are providers. While free sex within the movement was freely practised too, some materials (e.g., the Mo Letter 'The Old Church and the New Church') suggest that women were less willing than men to participate. This in itself is not problematic if the choice is possible. While theoretically this was the case (as seen above), in reality much pressure was put on women to submit. It is a pressure that emanates not just from peers, the movement or Berg, it has its origin in the divine.

In 'Revolutionary Women' Berg encouraged women to 'wear as little clothing as possible, so as to both partially reveal and yet at the same time partially and provocatively conceal her natural beauty and charm' (see Wallis, 1997, p. 13, see also Williams, 1998, p. 76). The sexual subjugation of women was not always problematic it seems. Williams remarks, 'What I did was in love and for love, and I think that faith is what protected me from the horrors and degradation that I witnessed in all the highclass call girls whom I met during that period of time' (1998, p. 44). Having left the group, however, Williams concludes, 'Clearly, the women of the cult had suffered greatly' (1998, p. 261). They were 'forced' to have babies, subjected to sexual and physical abuse, and stripped of their agency, especially with respect to sex (Boeri Williams, 2002, pp. 343–54). 'Children and women, although living in a "free-sex" environment, did not consider themselves free to choose' (2002, p. 354). Mercy, for example, had nine children in her 20 years in the movement.

Bainbridge, in his preface to Chancellor (2000) writes that the movement 'continues to endorse erotic sharing among adult members, so long as no one is harmed' (p. x). I do not want to suggest that the group set out to harm women. But what is endorsed and what happens are two separate things.

Irigaray, in 'Women on the Market' writes, 'The society we know, our own culture, is based upon the exchange of women' (1985, p. 170). The power that Berg had, as direct line to God, is at least partly about his authority to regulate the exchanges of all members.

> *Commodities thus share in the cult of the father, and never stop striving to resemble, to copy, the one who is his representative.* It is from that

> resemblance, from that imitation of what represents paternal authority, that commodities draw their value – for men.
>
> (Irigaray, 1985, p. 178)

In a hierarchy with Berg and God at the top, all members are gendered female. However, the finer grained hierarchy means that women generally, and those without positions of power, are gendered female in relation to their powerful male gendered (be they men or women) superiors. Irigaray describes society in a way that also captures sex and power within the Children of God: 'The law that orders our society is the exclusive valorization of men's needs/desires, of exchanges among men' (1985, p. 171).

In the 'society' of the Mo letters, the law that orders is the 'exclusive valorization' of one man's needs/desires. But because that man, Berg, is God's representative, this is seen as natural law. Kent argues that in the Children of God 'members misattribute divine authority to leaders whom they relate to emotionally as demanding parents' (2002, on-line). It is important to remember that the demanding parent is the divine. Kent points this out: 'By representing himself as God's mouthpiece, Berg was able to equate the traditional Christian virtue of "surrender to Jesus" with "surrender to Berg" ' (2002, on-line).

Conclusion

I don't want to suggest that sex is all The Children of God was about (or indeed all The Family is about). Peter Amsterdam, partner of the (now) leader Maria, comments:

> I know our views on sex are a real problem; I just don't understand why it is such a big deal. It is all we ever hear. We are mostly about Jesus, telling people about Jesus. Sex is only 2 percent of what we are. I just don't think it's fair to make that much of it.
>
> (Chancellor, 2000, p. 94)

It is not fair in another respect as well. The sexual politics of The Children of God don't seem to differ that much from contemporary Western society. Women are subjugated to male desire. The miracle is not that groups like The Children of God ever existed, but that it is so easy to mark them off from the rest of society. King notes that 'Social scientists have frequently pointed out that religious systems both reflect and reinforce cultural values and patterns of social organization' (1995, p. 15).

It shouldn't be surprising that we find similar dynamics in the rest of the world. Irigaray's work certainly deals with this. But perhaps analytically closer are arguments that pornography silences and subordinates women. Apropos the 'lively' illustrations, and the imperatives for women to be sexually available in the group, these arguments are another way of framing Berg's texts, even though 'Pornography is not always done with words' (Langton, 1993, p. 296). The ability for words (or other semiotics) to silence and subordinate relies on having the power to perform the required speech acts (Langton, 1993, p. 298). It is clear that Berg, as leader and divine, does.

While sex may only be a small part of the Children of God, it is certainly a site of silencing. The letters examined chastise, rebuke and silence not only words but actions. Given that sex has less to do with this subject position than power, it would not be surprising to hear from male ex-members that they had been subjected to similar controls. To be powerful in the Children of God, that is, to be heard, one needs to be properly aligned with Berg. To do this, one has to sacrifice the self. It is a choice that individuals should be allowed to make; but they should know that they are making it.

Notes

1 See ex-member accounts, especially 'It's My Life – Mercy's Story' at www.exfamily.org
2 For recent work on the group see Bozeman (1998), Kent (2002), Chancellor (2000), Boeri Williams (2002).
3 Thanks to Mercy for pointing this out to me. See also Chancellor (2000) and Bainbridge (2002). For a history of the Children of God, see Wallis (1979). Though there are some questions about Wallis's involvement with the group.
4 Whether Berg and Maria ever married is doubtful. Further, it is not clear that Berg's first wife (Eve) was ever divorced; hence 'mistress' and not 'wife'.
5 Reorganization Nationalization Revolution.
6 See Kent (1994) for a sexual history of Berg.
7 Religioustolerance.org
8 Flirty fishing (or FF-ing) usually involved young women using their sexuality to 'hook' new members. The press called them 'Hookers for Jesus', while Berg called them 'God's Whores'. See Barrett (1996), p. 113.
9 Religioustolerance.org. The site, <exfamily.org>, do report that they receive letters from people with questions about sexual relationships with current members. Thus FFing does still seem to be practised, albeit 'in a limited and somewhat clandestine way' (Mercy, personal communication).
10 Religioustolerance.org
11 www.exfamily.org
12 See www.thefamily.org
13 This brings special privileges. See Williams (1998).

14 Thanks to Catalyst London for the 'pleasure' of seeing some of these illustrations; also to material at www.exfamily.org. In keeping with pornography, the women are always 'young and beautiful' (Irigaray, 1985, p. 199).
15 Personal communication.

References

Bainbridge, W. S. (2002). *The endtime family: Children of God.* Albany: State University of New York Press.

Barrett, D. V. (1996). *Sects, 'cults' and alternative religions: A world survey and sourcebook.* London: Blandford.

Berg, D. (1973). Women in love. Mo letter 292. Acquired through personal contact via www.exfamily.org

Berg, D. (1978a). The girl who wouldn't. Mo letter 721. www.exfamily.org [accessed 23 March 2004].

Berg, D. (1978b). Happy birthday! – RNR rules – A compilation. Mo letter 663. Mo letters, vol. 5.

Boeri Williams, M. (2002). Women after the Utopia: The gendered lives of cult members. *Journal of Contemporary Ethnography* 31(3), pp. 323–60.

Bozeman, J. M. (1998). Field notes: The family/Children of God under the love charter. *Nova Religio: The Journal of Alternative and Emergent Religions* 2.1, pp. 126–31, 134–5.

Chancellor, J. D. (2000). *Life in the family: An oral history of the children of God.* New York: Syracuse University Press.

Enroth, R. (1977). *Youth, brainwashing and the extremist cults.* Exeter: The Pater Noster Press.

Irigaray, L. (1985). *This sex which is not one.* Ithaca, NY: Cornell University Press.

Kent, S. A. (1994). Lustful prophet: A psychosexual historical study of the children of God's leader, David Berg. *Cultic Studies Journal* 11(2), 135–88.

Kent, S. A. (2002). Misattribution and social control in the children of God. http://www.theonet.dk/spirituality/spirit97–10/cog.html [accessed 16 July 2004].

King, U. (Ed.). (1995). *Religion and gender.* Oxford: Basil Blackwell.

Langton, R. (1993). Speech acts and unspeakable acts. *Philosophy and Public Affairs* 22(4), pp. 292–330.

Mercy. (2000). It's my life – Mercy's story. www.exfamily.org [accessed March 2004].

Wallis, R. (1979). Observations on the children of God. *Sociological Review* 24, pp. 807–29.

Wallis, R. (1984). *The elementary forms of the new religious life.* London: Routledge & Kegan Paul.

Wallis, R. (1997 May/August). Moses David's sexy God. *New Humanist* 93, pp. 12–14.

Wangerin, R. (1993). *The children of God: A make believe revolution?* Westport, CT: Bergin and Garvey.

Williams, M. (1998). *Heaven's harlots: My fifteen years in a sex cult.* New York: Eagle Books.

Part III

Gender and Language Use in Religious Identity

10

'Restoring the Broken Image': The Language of Gender and Sexuality in an Ex-Gay Ministry*

Amy Peebles

The ex-gay community of practice

In this chapter I explore the ways in which religious beliefs about the nature of gender impact and affect both language and linguistic practice in an ex-gay ministry. Specifically, I demonstrate how the inextricable links made between sex, gender and sexuality within ex-gay evangelical theology allow transforming expressions of gender to be interpreted as transforming sexuality as well. While some of the changing gender expressions are predictably towards more of what could be considered a traditional cultural norm, interestingly there is also the creation and reception of a new freedom of gender expression within the ex-gay community, as concepts of masculinity and femininity are constructed that resist certain dominant cultural stereotypes and reframe what manhood and womanhood look like for these Christian men and women in particular.

Ex-gay individuals are defined here as self-identified evangelical Christians who have experienced or currently experience same-sex attraction, most of whom at one time understood themselves to be gay or lesbian, but who are now attempting to transform their sexual identity to conform with their understanding of traditional Christian theology and sexual ethics, which for them includes a moral conviction against homosexual practice. The term *ex-gay* can be problematic for a number of reasons, and numerous participants in my study expressed that they do not

* This research project was generously supported by a Social Science Research Council Sexuality Research Fellowship Program Dissertation grant, 2003–2004. I would also like to thank Keith Walters of UT-Austin for his supervision, as well as thank both him and Allyson Jule for comments on drafts of this chapter.

like being referred to by 'what they're not' or by 'what they used to be'; however, it is the media-popularized and commonly accepted term for this category of individual, and I use it here for ease of reference. While this research addresses a particularly controversial aspect of human sexuality, it represents one part of a larger study that deals with a plurality of possible narrative negotiations and resolutions as individuals deal with perceived conflicts between their religious and sexual identities. Thus, gay Christian and ex-ex-gay narratives are also part of the larger project (Peebles, 2004).

Eckert and McConnell-Ginet (p. 95) define a community of practice as an 'aggregate of people who come together around mutual engagement in some common endeavor. Ways of doing things, ways of talking, beliefs, values, power relations – in short, practices – emerge in the course of this mutual endeavor' (p. 95). They also state that the 'community of practice' concept helps researchers overcome reductionist assumptions common in language and gender research, such as genders being seen as independent of other aspects of social identity relations and an assumption that gender means the same across communities. The ex-gay ministry community of practice provides an excellent opportunity to note the ways in which understandings of both gender and sexual identity are reconstructed through language and practices emerging from 'shared belief and symbolic systems', especially because for ex-gay individuals, these understandings are primarily determined by their concurrent membership in another identity category: namely their religious identity as evangelical Christians.

In the summer of 2002, I spent three months as a participant observer at an ex-gay ministry, which I'll call 'Liberty'. During this time, I conducted ethnographic interviews and collected the 'life history' narratives of 20 men and 17 women who were either past or present ministry participants. I also attended and recorded a weekly women's Bible study and support group meeting. With respect to the men's weekly session, there was concern that the presence of a female researcher might inhibit some of the group members and keep them from feeling comfortable enough to share freely and openly. Thus, I did not attend the men's meetings, but arranged for and recorded several focus group discussions among smaller groups of the male participants. The data here are drawn from those corpora.

The ministry at Liberty offered what its creators described as a year long 'residential Christian discipleship programme' that focused on building the participants' individual relationships with God and helping them with their stated goal of dealing with the issue of same-sex attraction in their lives. There were two men's houses and one women's house, where between four and six ministry participants lived and shared household responsibilities, with a ministry staff person either present or available on-call at all times. It is important to note here that all participants were at Liberty of their own will and desire and were free to leave the programme at any time. Liberty has an 'if you see this as a problem in your life and want help, we're here' policy, does not recruit, and has an application and acceptance process. For example, in my study there was one woman who had been ambivalent about her participation, and upon her arrival, she expressed that it was mainly her parents who had wanted her to come. Ministry leaders then asked her to seriously reconsider, and though she was welcome to stay, she was encouraged not to participate if she herself did not truly wish to be there.

Liberty shares the distinctive features of a community of practice as delineated by Holmes and Meyerhoff (1999); namely group members share a set of required practices, define and construct membership internally, and have an active awareness of the interdependence of their personal and group identities. At the ministry, there was mutual engagement and regular interaction in that both men and women had separate study and support groups that met for two hours weekly, and members participated collectively in Bible studies and other ministry functions on other nights. Also, all members of the residential programme attended the same non-denominational evangelical Christian church. Additionally, group membership indicates participation in an active and situated social process of learning (Lave and Wenger, 1991) as the participants move towards the goal of sexual and spiritual identity transformation.

It should be noted that individuals in the ex-gay ministry seek a range of differing outcomes within that goal, such as no longer engaging in certain behaviours, no longer having proscribed desires, having heterosexual desires, or being in a heterosexual relationship, often depending on where they are in their 'journey' or 'process', as they would commonly phrase it. However, all shared the primary goal of 'growing in their relationship to God' and in obedience to their understanding of sexual morality; in fact, a key ministry slogan is: 'You are not here to overcome something; you are here to be overcome by Christ.'

Ex-gay theology and psychology

The religious beliefs concerning sex, gender and sexuality operative within Liberty, and ex-gay ministries more generally, are primarily drawn from their understanding of the Biblical account of creation found in Genesis Chapters 1–2. This account, along with procreation requiring one male and one female to reproduce human life, is interpreted as evidence that heterosexual partnerships are the 'created intent', which is a frequently used phrase in ex-gay ministries. The impact of this particular Christian worldview of gender and sexuality cannot be overstated, for in many ways it provides both the basis and motivation for the ex-gay identity transformation process and is alluded to or referenced in every ex-gay narrative I collected. For example, in Excerpt 1 below, Henry remarked on created intent and established it as a standard for 'living normally', thus implying a 'created norm' to which people should conform:[1]

(1) *Henry*: **I define normal as being created**, so when I attempt to do what I'm doing, when all of us attempt to do what we are trying to do, **changing our orientation, we are living normally, because we are trying to align ourselves with what we are created to be. God created me to be a man. I don't believe that God made people homosexuals**.

In Excerpt 1, Henry linked being 'created to be a man' with his belief that God did not 'make people homosexuals', a contrastive sequence emphasizing his belief that being 'made' a certain biological sex is inextricably bound up with the existence of an opposite sex as the only appropriate and intended direction for sexual expression. As with most religious and theistic traditions, Henry externalized agency to God, the Creator, as the One who established and designed what is and is not and with the power to ordain what should and should not be. However, in keeping with evangelical Christian theology, Henry retained individual moral agency by having a choice of response to God and His ways; in his life narrative, Henry framed his story as one of exercising this responsive moral agency by 'trying to align himself with what he is created to be'. Bart directly invoked the creation story and his understanding of it in his narrative, as in Excerpt 2:

(2) *Bart*: **I was born heterosexual**, and **I'm not in denial, that I've ever struggled**, but I was born a heterosexual. Christ called me from the very beginning, **God called me from the very beginning to reproduce with a woman, that He saw E-, Adam alone, and He sent Eve, a woman**.

In Excerpt 2, Bart stated that he was heterosexual from birth, thereby aligning himself with what he viewed as God's objective intent for him (i.e., 'God called me'), despite his subjective experience of same-sex attraction. This statement also indexed the ex-gay ministry tenet that homosexuality is not an innate or genetically encoded trait. Thus while

Bart acknowledged having 'struggled' with homosexuality, he chose to define his sexuality in terms of what he viewed as the intrinsic design of creation, which in his understanding is a potentially procreative union with a woman.

In terms of Critical Discourse (Fairclough, 1995), both Henry and Bart naturalized heterosexuality and biological sex via an oppositional denaturalization of homosexuality, thus reflecting their ideological worldview. A final example is given in Excerpt 3, where Deborah denaturalized her experience of same-sex attraction (i.e., 'anything that *seems* to be natural') in light of an objective 'way it's supposed to be' that she claimed emanates from the Creator:

(3) *Deborah*: Well, I would say to any-anyone who says I'm naturally attracted to women, anything that **seems to be natural, when God touches, it's the way it's supposed to be. I used to think that I used to be naturally born homo-homosexual, but God showed me something different, God showed me the right way to live. And when you know more about God, then you'll know more of who you are. And when you don't know God, you don't know yourself**. And so sometime we can get caught up in hearing what people say, hearing what-what we think, but if we don't know God, then we really don't know. **You got to know the Creator to know exactly who you are. And that was the problem-the issue with me.**

In Excerpt 3, Deborah claimed that only by 'knowing the Creator' could she know herself, and that she had to go to God to find out whom she was 'supposed to be'. Thus Deborah reported a relinquishment of her previous thoughts about being 'naturally born homosexual' and the authority to self-define, claiming to have sought the One who knew her true identity and would tell her 'the right way to live', which for her encompassed sexual identity, attraction and behaviour. Here Deborah demonstrated the ex-gay belief that there is a stable source of the self that originates outside the self, the Creator God, from whom the self and all aspects of identity are received and derived.

Andy Comiskey, a prominent ex-gay ministry leader and author of several books used at Liberty, clearly laid out the ex-gay theology relevant to the discussion here in the following quote, where both sex and gender are ontologically rooted in the being of the Creator and therefore have an origin and nature that transcend human beings. (I have bold-faced certain items in order to emphasize the recurrence of relevant 'terministic screens' within the text.[2])

> **God's intention for humanity** is represented by the **harmony of man and woman together**. But that freedom to be for another requires **security** in one's **personal identity as male or female**. Thus **gender security** matters profoundly. In paradise, that security was a given. But in the **post-garden reality of a child's development**, one can either **grow or fail to grow** into that confident posture. Whereas **biology determines one's physical sex, gender identity** involves the more complex **process of acquiring a sense of oneself as a male or female**. And that **process can go wrong**. Still, it remains true that **security** in one's own **identity as a man or a woman** precedes the freedom to be for another. The compelling nature of the **'otherness'** perceived in a member of the **opposite sex** results from the clarity and **security** one experiences in his or her own **gender identity. The image of God, then, involves gender identity *and* complementarity. God created gender in its duality as male and female**. And he **created** us as his representatives to discover that **duality**. In order to be true to the divine command, a person must reckon forthrightly and concretely with his **maleness or her femaleness in relation to the other**. The **'true self'** always includes one's **gender identity** and its **relation to the opposite sex**.
>
> (2003, pp. 25–6)

As seen in the above passage, ex-gay evangelicals accord ontological status to male and female because of their belief that the human dimorphism of biological sex was intentionally designed by God to reflect His image.[3] With respect to sex and gender, while these variables are not entirely conflated, they are inextricably bound together in an implicational relationship that involves a one-to-one mapping of biological sex on to the traditional correspondent gender and gender on to heterosexuality. In addition, there is a binaristic understanding of gender, where men and women are seen as 'dual' and 'complementary', having intrinsic and fundamental differences; relatedly there is a heterocentric understanding of sexuality that inherently links sexuality with biological maleness and femaleness and an intended pairing with the 'other', as in the quote above where one sex is never mentioned without a collocation or reference to a connection with its 'opposite'.

Comiskey's comments also reveal ex-gay beliefs about the psychology of gender and sexuality. Whereas a biological determinism is expressed concerning physical sex, gender identity is seen to involve a 'more complex process' of development and coming to a secure 'sense of oneself' that corresponds with being male or female. 'The post-garden reality of a child's development' alludes to the Biblical story of Adam and Eve's fall into sin, loss of perfection and subsequent expulsion from the garden; in short, ex-gays believe that all now live in a 'fallen world', one that is marred by sin and in which children can either 'grow or fail to grow' into a 'secure gender identity' and the development process 'can go wrong'.

Crucially, the attainment of a 'secure gender identity' is seen as directly related to the development of what ex-gays would consider healthy heterosexuality; as above, 'gender security' is said to precede the ability to perceive the 'compelling nature' of the 'otherness' of the opposite sex. Hence the ex-gay worldview characterizes homosexuality as a result of 'gender confusion' and most often attributes same-sex attraction to some type of disturbance in the process of child development. Thus a ministry newsletter described Liberty as 'a ministry for the man or woman who is gender confused, has same-sex attraction, and wants help due to the call of Christ in their lives'. Reflecting all of the beliefs described above, 'healing homosexuality' has commonly been referred to as 'restoring the broken image' (Payne, 1996a).[4]

The language of gender in ex-gay discourse

As with any worldview, the ex-gay belief system concerning gender and sexuality affects the language of its holders in profound ways. Consider Excerpt 4 from Ranni, where she described aspects of her behaviour when she was self-identifying as lesbian and involved in same-sex relationships:

(4) *Ranni*: Oh yeah, I wore only black, I was really into **the scene**, you know, I had my hair spiked and I'd write really bad introspective poetry. I would drink my coffee only black and I'd smoke imported cigarettes, and I was **acting out** what, if you even look at right now in the gay **lifestyle**, especially in the lesbian **lifestyle**, you-you see a lot of these **women acting out their false masculine**, and it was really **tough**, and it was a **refusal to be** anything quote-unquote **that a woman is** in the eyes of society or in the eyes of God because really it's the eyes of God because society reflects that. Some of it's good and some of it's bad, some of it is caricatured and stereotyped; it's not a full understanding of

> it. But, you know, **a woman is a woman** because that's **who God created her** to be. And I think you could see that across the cultural. It has nothing to do with how Western society mirrors it. There are some things that a woman is, period, and in **dressing up** the way I did and in **acting out** the way I did, I was basically saying 'no' to all of that – of-**to being a woman** because I **refused** to be victimized, and I **refused to be like my mother**, weak and passive and indecisive and paranoid, fearful, and uh emotionally a train wreck, and I couldn't go there.

In Excerpt 4, Ranni's language clearly reflects the ex-gay Christian worldview concerning both gender and sexuality. First, phrases such as 'the scene,' 'lifestyle,' 'acting out' and 'false masculine' invoke images of performance, of an enactment of an identity or a role that is a false characterization rather than an emanation from her true identity. Notably in this section of her narrative, Ranni's description is not made in terms of sexual involvement, but gender comportment, as she referred to the semiotics and style markers of her self-expression at that time, such as stereotypical 'tough' behaviours (i.e., drinking coffee only black, smoking imported cigarettes) and dress styles (i.e., a 'spiked' hairstyle and wearing only black).

In characterizing her behavior as a 'refusal to be anything quote-unquote that a woman is', Ranni set up both a volitional choice and an oppositional contrast. First, Ranni described herself as having chosen to 'say "no" ... to being a woman', which implied a volitional rejection of her true gender identity. Second, Ranni stated that her 'dressing up' and 'acting out' was in opposition to 'some of the things that a woman is – period'; thus while stating that there can be societal distortions of what women are (e.g. 'some of it's good and some of it's bad'), Ranni indicated her belief that there are some characteristics of a woman that are essential and that transcend culture based on 'who God created her to be'. Thus, with a twist on Butler's (1990) notion of performativity, Ranni viewed her gender identity as having an essential reality, and then discussed both her performance of gender and expression of sexuality in terms of whether she was acting in accordance with, or in opposition to, this reality.

Finally, Ranni linked her 'refusal to be a woman' with a 'refusal' to be like her mother and enumerated several negative traits and characteristics (e.g., weak, passive, indecisive) of her mother with which she would not identify ('I couldn't go there'). In so doing Ranni referenced one of the most common themes within ex-gay narratives; namely the etiology of same-sex attraction in their lives. With respect to developmental influences, the failure to bond or identify with the same-sex parent is seen to inhibit attaining a 'secure gender identity' and is frequently discussed within ex-gay ministries as a partial contributor to the development of homosexuality. For example, in *Healing Homosexuality*, a book studied at Liberty, Leanne Payne (1996a, p. 61) stated: 'For a young boy to seriously reject his own father (even with "good reason") is often to find that, as an adult, he has rejected his own masculinity' (from a chapter entitled 'The Problem of Gender Identity'). Beth directly referred to 'root issues' involving the same-sex parent and her ensuing gender identity at the beginning of her narrative, as in Excerpt 5:

(5) *Beth*: I just said, 'I give it all to You [God], including my **sexuality** and my **confusion** with my sexuality'. And, of course, that opened up the door then to get an introduction to Exodus [a large support and referral organization of affiliated ex-gay ministries] and the support group and to get now an understanding of those **root issues** from Beth Smith that **resulted in** the **choices**, whether consciously or unconsciously, that I made to go into the **lifestyle** and pursue the **behaviours** that I pursued, you know, and lived out. So with that then, with the Exodus ministry and the support group, coming to see that the **root issues** that I have is one, a **lack of bonding with the same-sex parent**.

In Excerpt 5, Beth discussed 'confusion' with respect to her sexuality and the process of getting involved in an ex-gay ministry after her decision to 'give it all' to God. Beth framed her years of same-sex relationships and lesbian identification as 'choices', 'behaviours', and a 'lifestyle' that 'resulted' from 'root issues', once again denaturalizing homosexuality and describing a perceived disruption in childhood development (i.e., 'lack of bonding with the same-sex parent'). From this segment, Beth went on to describe how she was 'more kind of like the boy' and discussed primarily participating in activities as a child that are stereotypically associated with male children. Thus again, Beth's narrative clearly demonstrated the one-to-one mapping of sex on to gender and healthy gender development leading to an opposite directed sexuality within ex-gay ideology.

Sociologist Christy Ponticelli (1993) described the ex-gay belief in God as the perfect parent and Father who can heal traumas and make up for deficits experienced in childhood to bring people into their true identities, including gender identity (see also Erzen, 2002). This belief is reflected in Payne (1996a): 'A man, unaffirmed in his masculinity, can fully integrate with it as he learns to come into the Presence of God, the Father, the Master Affirmer. There, listening to Him, he begins to "taste", as it were, the divine Masculine that resurrects his own' (p. 58). Thus again gender is believed to be transcendent and can be called out by the God of whose image it is a reflection. And due to ex-gay beliefs about sexuality, there is an implicational relationship that to receive affirmation in their gender will have an effect on ex-gays' sexuality as well. Consider Ranni's post-narrative interview, during which she stated that she no longer experiences same-sex attraction to any degree. In a follow-up question about emotional attraction, she claimed again that she had been 'healed,' as in Excerpt 6:

(6) *Amy Peebles*: Do you ever have to watch yourself with emotional intimacy with women?
Ranni: No. I don't anymore. That part has been healed. I don't and that's amazing. That stemmed from not being affirmed by my own mother, I think, and having that area affirmed in my life through the Lord has really helped me.

Here again Ranni related aspects of her struggle with women to 'not being affirmed' by her mother, and claimed that receiving God's affirmation in those areas 'really helped' and led to 'that part' being 'healed'.

Transforming the linguistic practice of gender

Ex-gay discourses often describe a process of accepting one's sex and gender in the journey of coming out of homosexuality; hence the linguistics of self-reference is often described as changing with respect to gender identification from a more neuter position to one embracing either male or femaleness uniquely. For example, the women at Liberty studied a book entitled *Out of Egypt: Leaving Lesbianism Behind,* in which the author recounted such a process and concluded with the statement: 'No longer would I refer to myself as a Christian person, as I had for the past several years. From now on, out of obedience to God, I would call myself a Christian *woman*' (Howard, 1991, p. 178).

In Excerpt 7, Justin described his process of laying aside a gay identity in the gendered terms of learning 'how to act like a man'; similar to previous examples he associated his manhood with God's creative action, to which he was then responding. Also, at 40 years old, Justin's statement of 'I haven't learned that yet' implied that there had been some disruption in the process of his male identity development and maturation:

(7) *Justin*: I'm a man, God created me a man, and I'm going to learn how to act like a man. I haven't learned that yet. That's what I'm doing. You can say I'm gay, acting like a man. But that's not what I'm saying, you have your perception, and I'll have mine.

Thus, due to the intertwined understanding of gender and sexuality in ex-gay theology, transforming expressions of gender can be interpreted as evidence of transforming sexuality as well. Consider the following from Beth's narrative in Excerpt 8:

(8) *Beth*: I was learning not to identify with 'gay', but I didn't see myself as a Christian woman [laughs], and probably at that time as a *woman* maybe, you know. I really, it was like I couldn't put a label on it. I was in that wilderness in my own personal identity, and that was probably part of that growing and that painful process, because I was just kind of – I don't want to say an 'it', but I didn't, you know?

In Excerpt 8, Beth described going through a 'wilderness' (as with Howard's book above, a Biblical allusion to Israel's journey out of slavery in Egypt) in her personal identity where she did not identify as a 'woman' and yet did not know what to identify as, being in the process of 'learning' to relinquish a gay identity. Notice once again that Beth set up an opposition between 'gay' and 'woman', not 'gay' and straight, indexing her belief that acceptance of a certain gender identity implicated a non-gay sexuality. Beth went on to describe a process of becoming more comfortable with herself as a woman and spoke of changes in hairstyle and dress that took place slowly over time; with respect to these external gender expressions, Beth said: 'When the healing happens on the inside, you'll see the reality on the outside.'

Not surprisingly, many of these transformations move towards what would be considered a more traditional gender expression. However, within ex-gay discourse there is often a discussion of receiving a new freedom of gender expression that does not necessarily have to correspond with societal or cultural definitions of what a man or woman *should* look like, as in Excerpt 9:

(9) *Beth*: Sometimes I find it hard to relate to a woman who's very, maybe exudes the feminine as prissy or something, because

I'm **not that kind of woman**, but I've come to a place in my life at 48 and 49 that **I'm content with who I am, and as long as I'm continuing to grow in *Him*, and I'm the person and the woman that *He* made me to be, that I'd be content with that**, because for a long time, **I've tried to be what everybody else wanted me to be** – and so, that's that **performance**, and, and you can perform and perform and perform, but you'll never make everybody happy and satisfy everybody. For me, **I'm content with who I am, and I'm not the quote unquote very feminine – I'm not the woman who walks on air, you know**. I'm a little bit more – uhn, uhn, uhn [claps hands 3 times in sequence with short, abrupt sounds], [laughs].

Thus, in Excerpt 9, Beth contrasted herself with a 'prissy' or 'very feminine' woman, stating that she was 'not that kind of woman'. She claimed a contentment and freedom with her gender identity with respect to not living up to certain 'feminine' standards or stereotypes and stopping trying to 'perform' for everybody else to be what they 'wanted her to be'. Again she referenced identity and affirmation as coming from God, and stated that as long as she was 'the woman that He made her to be', she would be 'content with that'.

During a men's focus group discussion, I asked participants about their concept of masculinity and where it came from. This led to the following interchange in Excerpt 10, in which the group members set up a contrast between what they viewed as contending concepts of manhood; namely masculinity according to God versus the 'the world':

(10) *Bart*: **I feel that God needs to show me who He wants me to be.** And His Holy Spirit needs to convict me and show me. **I want Him, through His Holy Spirit to show me what masculinity is, you know**? Because the world says a man should be sittin' back, you know, holdin' his crotch, watchin' TV, you know, you look at, what's the guy's name that, you know? Watch TV, ???

Simon: 'Married with Children.'/

Bart: ['Married with Children'] you know what I'm saying? He sits there in back [with his hand in his pants],

Jon: [that is *no:t* a good]/

Bart: /drinkin' a beer, watchin' pop porn, [you know what I'm sayin'?]

Simon: [that's my dad].

Bart: or, you know, and/
Simon: /that's my role model/
Bart: /bein' rude to the mo-you know, the wife, or you know [just bein'] real passive.
Simon: [Controlling.]
Bart: You know, I mean just *all* these different things, you know, that the world does./
Simon: /[Degrading.]
Bart: [And then] then the worl-the-a male watches football. A male does this, a male does that./
Simon: /Yeah./
Bart: /Not necessarily. [You know what I'm saying? A male] does not have to be a
Simon: [Not every single one.]
Bart: 'hrooo hooo hooo.' [Neanderthal-ish sounds]
Simon: And you're made to feel if you don't, then you're not a man.

As Excerpt 10 shows, Bart first stated that he wanted 'God to show him' what masculinity is. He then began describing what 'the world says' a man should be and illustrated this via a negative TV image of manhood from the sitcom *Married with Children*. Bart listed several features of the stereotypical insensitive, boorish male and described the male protagonist on the show as a beer-drinking, TV-engrossed, passive, rude man 'sitting back holding his crotch' and watching 'pop porn'.

After Jon agreed that the TV show and image was 'not a good' thing, an interesting sequence occurred where Simon began a cooperative series of latches and overlaps to both agree with Bart and identify the image with his own father. Simon's statements of 'that's my dad', 'that's my role model' are a reminder of the ex-gay belief in the importance of gender identification with the same-sex parent, but the 'worldly' image being constructed here is negative and one with which Simon clearly did not wish to identify and did not value. Simon's overlapping negative attributions of adjectives such as 'controlling' and 'degrading' further indict this construction of masculinity.

Bart then vied against a monolithic concept of manhood by referring to stereotypes such as 'a male watches football' that are 'not necessarily' the case, which Simon agreed with by his overlap of 'not every single one'. The interchange climaxed with Bart's statement that 'a male does not have to be' followed by his vocalization of ape-like, grunting noises, thereby performing the Neanderthalish stereotype he had been constructing from the TV show. Simon ended with the critique that if one doesn't fit certain stereotypes (e.g. watching football), 'you're made to feel like you're not a man'.

Thus, in Excerpt 10, Bart and Simon collaborated and displayed several key beliefs about gender from an ex-gay perspective. Namely these men believe that while there are different concepts of manhood according to 'the world', many of them are undesirable, distorted and not what God intended. Bart believed that God could reveal to him 'who He wanted him to be' and 'what masculinity is', and similar to Beth's narrative, this masculinity does not necessarily correspond with or live up to certain cultural stereotypes.

A final example of the transformation of linguistic practices related to gender and sexuality comes from the same focus group session above. At Liberty, there is a ministry rule that prohibits camping, a stereotypical gay male speech performance, among the ministry participants. The men's discussion of this rule is given in Excerpt 11, excerpted greatly due to space limitations:

(11) *Bart*: Well, **why is camping not allowed in the programme** I just feel is that, um, it's just because of the fact that **it's glorifying the *old* man** ...

Jon: It does glorify the old man. And that's not why we're here. **I think it's easy to have the mannerisms and talk like we used to talk in the lifestyle, but that's what we're growing out of** ...

In Excerpt 11, Bart succinctly explained the reason for the rule by identifying camping with 'the old man',[5] a phrase used frequently in the New Testament as a metaphor for the self that is associated with sinful behaviour and attitudes prior to conversion to Christianity. Jon agreed with Bart and discussed a process of 'growing out of' certain mannerisms and speech styles that were from 'the lifestyle', some of which were described in the discussion as being inappropriately feminine. The men thus cooperatively encouraged one another to lay aside camping as part of their 'old ways', and as Mel later put it, to 'start acting like men', thus again fluidly linking gender expression with sexuality.

Conclusion

In conclusion, I hope to have demonstrated how the ex-gay ministry functions as a community of practice in which both sex and gender are granted elements of an essential and transcendent status. Due to the direct implicational relationship accorded between sex, gender and heterosexuality in the ex-gay Christian worldview, the modification of gender expression can be interpreted as evidence of change and healing with respect to one's sexuality as well. While shifts towards more traditional cultural norms of femininity and masculinity often occur, interestingly there is a significant resistance to certain cultural stereotypes as well. Ex-gay men and women search for and discover alternate constructions of heterosexual masculinity and femininity, which in some ways approximate cultural norms and in other ways directly resist or repudiate them.

The ex-gay ministry becomes a community of practice in which participants encourage each other with respect to their linguistic enactments of both gender and sexuality, all flowing from what they believe to be an ontological truth about their identities as men and women made in the image of God. Thus religious beliefs about gender profoundly affect the way ex-gay individuals both talk about and live their lives, especially as they pursue a largely theologically motivated sexual and spiritual identity transformation.

Notes

1 All names are pseudonyms.
2 Kenneth Burke (1966) developed a notion of 'terministic screens', whereby one's choice of terms acts as a screen or filter. Once the terms are in place, the resultant 'filter' directs the attention of speakers in particular ways, thereby affecting their observations and perceptions of reality.
3 This belief is based on their interpretation of Genesis 1: 26–27: 'God created man in His own image, in the image of God He created him; male and female He created them' (New American Standard Bible).
4 Note that evangelicals believe all humans are 'broken' from sin. 'Homosexual behavior is merely one of the twisted paths this fallen condition in man takes. ... We are all fallen, and until we find ourselves in Him, we thrust about for identity in the creature, the created' (Payne, 1996a, p. 125).
5 For example, 'Do not lie to one another, since you have put off the *old man* with his deeds', Colossians 3:9 (New King James Version).

References

Burke, K. (1966). *Language as symbolic action: Essays on life, literature, and method.* Berkeley: University of California Press.

Butler, J. (1990). *Gender trouble: Feminism and the subversion of identity.* New York: Routledge.

Comiskey, A. (2003). *Strength in weakness: Healing sexual and relational brokenness.* Downers Grove, IL: InterVarsity Press.

Eckert, P., and McConnell-Ginet, S. (1992). Communities of practice: Where language, gender, and power all live. In: K. Hall, M. Bucholtz and B. Moonwomon (Eds), *Locating power: Proceedings of the 2nd Berkeley women and language conference*, Vol. 1. Berkeley, CA: Berkeley Women and Language Group, University of California (pp. 89–99).

Erzon, T. (2002). *Out of Exodus: The ex-gay movement and the transformation of the Christian Right*. PhD dissertation. New York University.

Fairclough, N. (1995). *Critical discourse analysis: The critical study of language*. Boston: Addison-Wesley.

Holmes, J., and Meyerhoff, M. (1999). The community of practice: Theories and methodologies in language and gender research. *Language in Society* 28, 173–83.

Howard, J. (1991). *Out of Egypt: Leaving lesbianism behind*. London: Monarch Books.

Lave, J., and Wenger, E. (1991). *Situated learning: Legitimate peripheral participation*. Cambridge and New York: Cambridge University Press.

Payne, L. (1996a). *Healing homosexuality*. Grand Rapids, MI: Hamewith Books, Baker House.

Payne, L. (1996b). *The broken image: Restoring personal wholeness through healing prayer*. Grand Rapids, MI: Hamewith Books, Baker House. (Original work published 1981.)

Peebles, A. (2004). *Sexual and spiritual identity transformation among ex-gays and ex-ex-gays: Narrating a new self*. PhD dissertation, University of Texas at Austin.

Ponticelli, C. (1993). *Fundamentalist ex-gay ministries: A loving approach to a sinful problem*. PhD dissertation. University of California, Santa Cruz.

11

'*Assalam u Alaikum*. Brother I have a Right to My Opinion on This': British Islamic Women Assert Their Positions in Virtual Space

Fazila Bhimji

This chapter explores linguistic practices of second- and third-generation young Muslim women in a specific context: an Islamic on-line community based in Britain. This particular chapter is part of a larger ongoing study of British Islamic women's identities in multiple spheres such as Islamic study circles, Islamic Magazines, Public Speech and Television Documentaries. This study examines particular linguistic practices of Muslim women who participate in discussion threads along with Muslim young men. The study will demonstrate that these young women argue and debate with other on-line participants, contest mainstream notions and depictions of Islam, and display their knowledge during on-line discussions. In doing so the study aims to contribute to the theoretical discussions on language and gender, where gendered identities are conceptualized in ways that do not always set women apart from men. Furthermore, the study shows that even women who express themselves in Islamic ways can have varied identities such that they can be religiously inclined and assertive in the same instance. Additionally this chapter focuses on commonsense understandings of Muslim women as passive, subordinate and having limited access to knowledge.

As women's agency, choice and voice have replaced earlier feminist preoccupations with passivity, oppression and silence (Bucholtz, Liang and Sutton, 1999), this study draws on what is known as feminisms' second wave. There has been much scholarship on language and gender, and there has been substantial interest in second- and third-generation Muslims in the United Kingdom. However, little attention has been given to the ways in which Muslim women use discursive forms to subvert beliefs which position the women in subordinate positions.

There has been much focus on Islam and Muslims in scholarship in recent years. Given the attention given to Islam post-9/11 and the subsequent rise in Islamophobia, several scholars have devoted much space to explaining the diversity of Muslims (in regard to ethnicity, religious and political practice) residing in the West as well as Muslims living in Asia and the Middle East. Given that Islam has been perceived in the mainstream as a religion which restricts women's rights, many scholars have sought to challenge these historical and modern stereotypes. For example Bullock (2003) challenges the Orientalist notion of Islam where Islam is constructed as barbaric, violent, medieval and backward. In her study, Bullock relying on in-depth interview data and testimonials of some 15 Muslim women, presents an alternative perspective, where women do not always view the 'veil as oppressive', but rather perceive it to be liberating from a capitalist culture dependent upon cosmetics and the ideal thin women. Pnina Werbner (2002) in her ethnographic work with the Pakistani Muslim Diaspora in Northern England contests Orientalist perceptions of Islam as she discusses Pakistani/Islamic/Mancunian/British/South-Asian/Punjabi women's participation in a wide range of spaces which she defines as simultaneously *public and familial, Islamic and culturally open*. She traces Islamic women's participation in the public sphere, where the public sphere is defined as a series of interconnected spaces in which the pleasures and predicaments of Diaspora are celebrated and debated (p. 15). While these scholars examine the role of Muslim women within a national and institutional framework, where there are less limitations, others have elected to examine women's roles in contexts where there are greater state impositions on expressions of Islamization. For example, Raudvere (1998) examines the resurgence of Islam among urban women in Turkey, where women challenge facts concerning secularism, religion and modernity. Other scholars have examined the ways in which participation in religious life acquires political dimensions (e.g. Kamalkhani, 1998; Mahmood, 2003; Gole, 2003). Gole (2003) in her study of identity formation of Muslim women argues that a transformation from Muslim to Islamite is the work of a collective counter-cultural movement. Because Islam is no longer transmitted by their social, family and local settings, these Muslims reappropriate, revisit and re-imagine collectively a new religious self in modern contexts (p. 815). However, little attention has been given to the linguistic practices of these social actors as they seek not only to challenge existing stereotypes but also struggle for their political rights. It becomes particularly crucial to examine linguist features when asking questions concerning identity formation. A study of linguistic resources can provide an important focal lens in comprehending the ways in which individual speakers present their individual identity and negotiate their social roles within a large community.

Gender and language

Much scholarship on language and gender has focused on examining gendered features such as markers of cooperativeness and interactional supportiveness. However, recent scholarship has shifted from presenting limited descriptions of gendered speech and style. More recently scholars (e.g. Goodwin, 1990, 1999; Mendoza-Denton, 1999; Orellana, 1999) debunk the myth of female submissiveness and show ways in which girls use strong language in varied naturally occurring contexts and texts. For example Goodwin (1999) demonstrates that Latina girls during play use oppositional stances to assert their respective position. Similarly Orellana (1999) shows how Latina/Latino student writers construct brave and even 'bad' selves through stories. Mendoza-Denton points out the ways in which the turn initial *no* can have properties of strict semantic negation, mark oppositional stance, and create certain elements of collaboration.

In a similar vein, while examining discursive practices of Islamic women such as rhetorical questions, strong assertions and conflict talk, this study will join researchers who conceptualize gender in complex ways. As Holmes and Meyerhoff (2003) suggest, these researchers understand gender identity as a social construct rather than a 'given' or 'fixed' social category to which people are assigned, where gender in their view is understood as the way individuals 'do' or 'perform' their gender identity through interaction with others, while an emphasis is placed on various aspects of interaction. More importantly, these scholars point out that not only do people speak differently in different social contexts, as sociolinguistic analyses of different styles have demonstrated. More radically, *talk itself* actively creates different styles and constructs different social contexts and social identities as it proceeds. Hence such understandings of gender, language and aspects of identity offer alternative ways of understanding gendered identities that go beyond the essentialist notion that gendered identities are inevitable, natural and fixed.

On-line Islamic groups

In recent times there has been much activity, discussions and debates on-line about a wide range of topics pertaining to everyday Islamic practices, Islam and citizenship, Islam and identity, and the current state of Islamophobia in the West.

There are over 100 Islamic groups on-line, which serve as a site for both men and women to articulate their positions on many of these ongoing debates. Many of these on-line communities tend to be geographically based, while others tend to cross national boundaries. Both men and women participate in these discussions. As the following paragraphs will illustrate, much of the time female participants are equally vocal in these virtual spaces.

A growing body of research has found that females and males display different participation patterns on-line, where females tend to participate less and receive less attention than males in mixed-sex on-line discussions (Herring, 1993, 1996). However, much of this research has concentrated on examining filtration behaviour (e.g. Panyametheekul and Herring, 2003) and on-line discussions where current issues/debates have yet to be examined. In on-line Islamic communities, I found that much of the time, as I mentioned above, the young men and women discussed a wide variety of topics where Islamic women were equally (if not more in certain cases) forceful in controlling the conversational floor. The following paragraphs illustrate some of the participation patterns and discourse styles found among Islamic-British chat-room participants.

Some features of computer mediated conversations

When speakers are on-line they do not have the advantages of face-to-face conversation. They cannot, for example, indicate through sustained eye contact where turn changes are likely to occur. In order to circumvent the coherence problems caused by lack of non-verbal cues and disrupted forms of turn-taking, participants use various means such as addressing others by name and engaging in conversations with the group at large rather than targeted individuals. According to Panyametheekul and Herring (2003) turn-allocation behaviour during on-line conversations is not necessarily different from ordinary conversations, since participants can self-select their turns at any time by adhering to a variety of strategies. Similar to ordinary conversations, issues of social appropriateness remains at the forefront and are determined in part by speaker identities and roles. Panyametheekul and Herring found systematic differences in participation patterns and discourse styles of males where males tended to dominate in amount and manner of communication, using confrontational and self-promotional talk while females tended to be attenuated, self-deprecating, and supportive of others.

Some of the features of gender differences as pointed out by Panyametheekul and Herring include males posting longer and more messages, using assertive language, rhetorical questions, showing opposed orientations and using evaluative judgments; and conversely for females, posting shorter messages and receiving fewer messages, using hedges and qualifiers, and showing support and agreement with others. This chapter examines the use of discourse styles among Islamic women and demonstrates that the gender lines are not so neatly divided and that women also show opposed orientations, use assertive language, post long messages and use rhetorical language.

Data and methodology

The data were collected from the website 'The Revival'.[1] The Revival describes itself as an Islamic on-line youth group concerned with issues facing the Muslims of today from sex, drugs and rock' n' roll to politics, arranged marriages and alien nations. The website was formed in 2001. Although, there are a number of websites where young people participate to discuss a wide range of issues, I selected this site because it was UK-based and I am interested in examining the ways British Islamic Women articulate their language and identity. Many of the other sites were based in the United States or were worldwide. Moreover, the site appeared to be very active and popular and the range of topics explored was very wide. Some of the titles include: 'hijab and jeans', 'ban on hijab', 'head-scarf hypocrisy', 'women's prayer', 'why no female prophets?', 'racism in the UK', 'British or Muslim how does it differ?', 'What is our identity?'. Many of these topics pertained to young women growing up in Britain and I was interested in examining the ways in which the young women participated in these conversations.

As a South Asian (secular) Muslim I was in a position to recognize many of the Arabic names and terms employed by the on-line participants. Hence I was in a position to identify the female and male participants based on their names as well as some of the nicknames they used. Moreover the participants often used Arabic terminology which I was familiar with since I had been raised Muslim in Pakistan. There were some cases where I was not familiar with the Arabic terminology that the participants used. In such instances, I asked some of my Muslim students at the University (where I am currently teaching) in Britain to translate for me.

Conveying Islamic identities on-line

While the veiling is explicitly symbolic of an Islamic identity, the question then arises as to what ways the young women convey their Islamic identity on the web during discussion threads. A close examination of on-line discussion groups show that one of the principal ways that young men and women express their identity is through the use of Islamic/Arabic language. The young men and women in their greetings employ the Islamic/Arabic greeting: *Assalam u Alaikum* (Peace be on you) *Assalam u Alaikum wr wb* (Peace be on you as well) *Salam* to Muslims (Peace Muslims), 'Peace' to Non-Muslims, and *Assalaam u'alaikum wa rahmutullah* (Peace be on you as well). These greetings clearly explicate an Islamic identity since the more common forms of greetings in Britain include 'Hiya', 'Hey', 'Hi there', and simply 'Hi'. I did not come across any of these greetings in the on-line texts. Similarly, closings include Arabic words such as *Wa Salaam* (Peace). In certain cases, many of the participants select to employ the Arabic word *Jazakallah* (May Allah Reward You) rather than 'Thank you'. One practicing Muslim in Manchester informed me that even his 8-year-old niece insists on using the word *Jazakallah* as symbolic of an Islamic identity whereas the English 'Thank you' is perceived to be used by non-Muslims. This is particularly interesting when South Asian Muslims use this term because South Asian Muslims frequently use the Arabic greeting *Assalam u Alaikum* particularly to greet elders and the term *Jazakallah* is seldom used to express gratitude. In fact, it was among British Muslims that I first heard the term and it was explained to me that British Muslims purposefully elect to use this Arabic word to express their Islamic identity and that many believed that by employing certain Arabic words in place of English words they were following the Sunnah (The prophet's lifestyle). In addition to employing Arabic words in their greetings and closures, on-line participants also used Arabic words in the midst of their texts. For instance, the word *Alhamdulilla* (with God's blessing) was frequently used in texts. The use of the word *Alhamdulilla* is commonly used among Muslims in order to convey gratitude to God.

Apart from the use of Arabic/Islamic words in texts, in many instances the young men and women elect to self-identify as Muslims by using icons such as mosques or Islamic calligraphy alongside their names. In certain instances, where the young men and women who do not always wish to self-identify themselves by their names often use terms which show that they are Muslims. Some of these self-identifications include names such as 'Shia Sister', 'Cat Stevens', 'Muslimah', 'Sister', Nikaabi.

It appears that young women often select to use the word 'sister' since Muslims often address each other using the kinship terms 'sister' and 'brother'. Although these Islamic/Arabic words occur frequently during on-line talk, much of the time the discussions are carried on in English and in many cases the language the young people use shows that they are young British English speakers. For example, the young men and women employ informal language and abbreviations which are frequently used in Text messaging in Britain such as *4* rather than 'for', *2* in place of 'to', *gr8* rather than 'great', *thgt* in place of 'thought', *y'all* rather than 'you all', *tho* instead of 'though', *u* for 'you', *ur* for 'your'. However, it is particularly significant that the young people elect to employ a pan-Islamic identity and there is virtually no reference made to their ethnic identity. None of the participants use language or markers whereby one could identify their ethnic identity. This Islamite identity can be understood as a form of 'new identity' for many young British Muslims where they may distance themselves from their ethnic identity in favour of a pan-religious identity. In doing so, the young men and women are in a position to widen their social networks and meet Muslims from varied backgrounds in different situations since they do not have to restrict their networks to their own ethnic community. In these on-line discussions the common themes that emerge revolve around Islam rather than ethnicity. For women, these discussion threads could become quite significant as it allows them to argue, debate, discuss and explore issues related to Islam. The following paragraphs will discuss the ways in which young Islamic women employ language in these on-line discussions.

Using assertive language

The discourse styles that Islamic women select to adopt during online discussion threads reflect their complex social worlds. The women as the following examples will illustrate initiate topics, argue, strongly disagree, use collaborative language, and assert their respective positions on a wide range of issues. As several scholars (e.g. Goodwin, 1990; Mendoza-Denton, 1999) have demonstrated, the linguistic devices that speakers select to use display their stance where 'stances reflect and construct aspects of social identity as speakers take up positions associated with particular social categories and groups' (Mendoza-Denton, 1999, p. 273). The following excerpts selected from an on-line discussion thread reveal that young women deploy linguistic devices (such as overt disagreement and disaffiliation) such that their choices reflect their multiple worlds and debunk the stereotypical characterizations of passive and silent Muslim women. In many instances, these young women use argumentative language which lends evidence to the fact that these young women can be assertive, articulate and Islamic at the same time. However, importantly, even as they display opposition, the young women use linguistic devices which exhibit their friendship towards one another thus revealing their multiple identities in their on-line interaction. The following segments show some of the ways in which they invoke argumentative talk:

(1) *Omrow*: Salaam Aliyah.
I don't think you know what the hell you are going to reply if someone asked you where you're from. Tell you what. Why don't you sit down. Touch the ground below you, and name it. Then remember that answer. It will help you answer that question.
Aliyah (Female): Assalaamu'alaikum brother Omrow
I found that kind of rude ... but I guess that is just how you are ... why would I say I'm from England if I'm not? I have different origins and was born in a different country ... you could consider England as my third country. And if somebody asked me where I was from, England would be the last answer that I would give!!
Wa Salaam
Aliyah.

(2) *Sister*: Re: Have Salafis taken over the Muslim world? No
Reply on: 26 Dec.
On 25 Dec., 2003, 11:33pm, NBZ wrote:
The wahaabi are deviated as they **are not willing to accept the truth**. They have discarded rules of Ahadith, and fiqh to reach new conclusions which are AGAINST opinions of earlier scholars!

And **you're willing to accept the truth?** Where's your proof that they've discarded rules of hadith?
They do their best to stick to Qu'ran and Sunnah as class as possible.
How many of you done that?

(3) *Sister*: Re: Have Salafis taken over the Muslim world?
On 26 Dec., 2003, 11:37am, Guest-Abu T wrote:
The above posts regarding ibn baz and uthaymin are in error.
They did err and did go against the jama'a

Abdulwahab's own brother and father wrote books refuting his misguidance. Note that islam-qa is not the most reliable website.

Where's your proof mister?
As far as islam-qa, I've done my research on that and sheikh of this site do not go against Qu'ran and Sunnah.

(4) *Ambarin* (Female): Re: Music Video
Assalamu Alaikum
Music ... such a waste of time.
Abu T (Male): why write provoking comments ambarin?
Music is permissible, especially the nasheed type and I find that if you do not waste time listening to it, there is no harm.
Ambarin (Female): Re: Music Video
Assalam u Alaikum
Brother with all due respect I have a right to my own opinions on this. I am not alone in not liking music. I did not mention and nasheeds did I? I was speaking about music. Nasheed with duff I would consider listening to, but anything other than that I disregard.
The reason why I do not like music is because apart from lyrics, instruments etc it can take away some of the beauty of hearing the Qu'ran recited. If a person listens to popular music all the time, hearing Qu'ran recitation without instruments would in comparison seem bland to them. Their heart would not be captured by it, and most likely after a few minutes they would lose interest. Also its human nature when you hear a track to hum it, as you walk or do your daily activities. I'd rather listen to Qu'ran and have that in my memory, rather than a few tunes that take me away from the remembrance of Allah.
So that is one of the reasons I dislike music, including nasheeds, I do not see how that disrespects you brother.
KM: ... if a person listens to popular music all the time, hearing Qu'ran recitation without instruments would in comparison seem bland to them ...
By contrast i would say that after listening to the blandness of music all day, people would rejoice by listening to the diversity of (qaari) Recitation of the Holy Qu'ran.

(5) *Ambarin* (Female): Re: Music Video
Assalamu Alaikum

> That is up to you, I have a right to my own thoughts. One cannot deny that a person will remember Allah more if he/she takes up their time in Islamic activities and subjects rather than non Islamic ones.
> I don't like music, and the best thing I ever did was destroy my cd's. I don't need such diversions.
> p.s. Why is it that ppl give up music for the ramadhan period? There must be a reason behind it.

In the above excerpts the young women voice their perspective in assertive ways. In Segment 1 when Omrow confronts Aliyah regarding her place of origin, Aliya responds immediately and asserts her position strongly. The shape of the turn shows that Aliya does not in anyway mitigate her position, but rather emphasizes her position by using a rhetorical question: 'Why would I say I am from England when I'm not?' Panyametheekul and Herring (2003) in their study of chat rooms observed that it was males who were far likelier to employ rhetorical questions and women generally used questions as a means to elicit responses. However, in this context, Aliyah employs a rhetorical question in order to display her position. Moreover, Aliyah continues the dialogue by justifying her opinion: 'I was born in a different country and have different origins.' Finally, she concludes by showing her affective reaction by using two exclamation marks.

Similarly in Segment 2, a female participant who identifies herself as 'sister' displays her position. In this segment she actively challenges the participant by repeating part of the talk that is being opposed:

NBZ: the wahaabi are deviated as they **are not willing to accept the truth**.
Sister: **And you're willing to accept the truth?**

As Goodwin (1990) points out, through partial repeats 'the speaker is able to caricature a prior speaker by portraying his/her actions as ridiculous or inappropriate and the current speaker is thus able to both build a small effigy of the party being opposed, and thus display his/her own affective alignment to the actions that such a person performs' (p. 147). Similarly in this case, the on-line participant conveys her strong opinion by repeating prior talk rather than simply disagreeing with something in prior talk.

In Segment 3 a female participant responds by asking a rhetorical question: 'where's your proof mister?' By employing a pejorative person descriptor 'mister' the speaker further emphasizes her oppositional position where she not only opposes the participant's prior talk but also the person who produces the talk.

In Segment 4, the speaker, Ambarin once again actively disagrees with the prior speaker, KM. But rather than simply confronting the participant she justifies her response, explaining her reasons for disliking music. Furthermore, when KM continues to disagree with Ambarin, Ambarin continues to strongly adhere to her position: 'I have the right to my own thoughts.' In doing so, Ambarin makes the prior speaker's talk insignificant since she displays that his views do not affect her own viewpoint. In other words, Ambarin manages to show that she remains unaffected by the prior participant's point of view and that she is in a place where she can make her own judgment.

Mixed versus single-sex exchange

The above examples showed the ways in which Muslim women expressed oppositional stances. More importantly, the young women articulated their viewpoints in mixed-sex exchanges. Much of the literature discussing language and gender (e.g. Goodwin, 1990; Mendoza-Denton, 1999) shows that young women often use strong language in single-sex settings. For example, Goodwin in her study of second-generation Central American girls at play argues that the girls use verbal as well as embodied gestures to display their stance during hopscotch games. In this instance, the women contest and argue with men in virtual space. The following examples illustrate this:

(6) *Oldham Dude*: Re: NO TOPIC

cigs are no good. but i got some sufis dudes in the area they use marijuana Smoke it

Dawud: Re: NO TOPIC

Assallamu Allaikum.

Oldham Dude,

You really need to get a grip on reality. Your posts are almost always offensive. The 'joke' you posted is just inane. If that is the kind of thing that you feel is funny and feel has a place on a Muslim site then you need to go and speak to your 'molvi' again. I'm sure that the Revival editor will remove that

angry inspector: Assalam-u-Alaikum

I have been on this forum for ages but I haven't ever posted on here but now I have lost my temper so I'm posting a complaint that you shouldn't delete because it's for the forums own good and I know your temper so please don't but if you do then you will keep this thread the same state ☹! Please get rid of that post and other offensive post's. This forum is getting worse by the minute and it is very unfair

Sis: I completely agree!! and I'm also very angry but I'm not as patient as 'angry inspector' because I'm leaving for good now!!

Wassalam

Segment 6 shows that a young man who self-identifies himself as Oldham Dude makes a derogatory remark about Sufis which in turn invokes angry reaction among other on-line participants. However, it is a woman who identifies herself as Muslim Sis who expresses most anger. She uses intensifier adverbs such as 'completely', 'very', followed by several exclamation marks and angry faces. Her tone is not only angry but there is also an element of finality to it since she threatens to leave the virtual site. Segment 7 illustrates another example of a mixed-sex exchange:

(7) *Shabana*: Re: NO TOPIC

How come men are allowed to marry up to 4 wifes; aint that a bit greedy? Whats wrong with one?

Salam

Shabana (Female): Omrow wrote:

Men can handle four women thats why.

nah …

Omrow wrote:

Muslim women cannot marry non-Muslims because children adopt the religion of the father

Hey, I thought if a child was born by a muslim mother, it is muslim …

Omrow wrote:

A Muslim woman would not wish her babies bowing to the Pope. No way. Therefore, its all for the future of children.

Yeah, I agree with you there, but this happens most of the time anyway, muslim man marries jewish/christian women, child grows up and dont want to be brought up as a muslim … and that does happen most of the time anyway …

In this example, Shabana initiates a question concerning polygamy in Islam. A male participant responds to the question (not shown in the text) and Shabana contests some of his positions by quoting portions of his argument. Note that she quotes him in a small font whereas employs a larger font when presenting her own viewpoint. By quoting the male participant in this manner she is able to highlight exactly the points she wishes to contest and thus show her own agency as a female contesting a male participant.

The segments above show the ways in which British Islamic women express oppositional orientations. They use linguistic devices such as rhetorical questions, strong disagreement, strong assertions and terms of address which show that Islamic women cannot be viewed in monolithic terms as subservient, female and Islamic.

However, these young women do not simply engage in argumentative talk with each other, they also use this virtual space to counter mainstream perceptions and representations of Islam. The following segments show some of the ways the young women achieve this:

(8) *Nikaabi* (Female): Re: Chirac: Hijab is oppression

Personally I think it's ridiculous ... what harm are these girls causing anyone by wearing their headscarf?? ... does this mean they will put a ban on other items of clothing like bandannas, caps etc? ... i think not ... why is the headscarf any more threatening that other clothes ... whatever happened to 'freedom of expression'?

interesting to note that the leaders of a country feel 'threatened' by a group of young girls . ...

Shezaadi (Female): Re: What is our Identity?

yep you guys are correct ... there was mostly women in niqaab! there was one particular scene in it showing girls in all white including niqaab, all which was revealed were their eyes and they were doing exercises ... but from the angle it was shown it seemed as if these girls were doing vigorous exercise for 'jihad' – and it was only P.E.!!!!!

As the segment above shows, the young women do not employ mitigated language in order to express their views, but rather assert their position in strong ways. Nikaabi, for instance, employs a number of rhetorical questions followed by an assertive response, 'I think not', which is then followed by a series of rhetorical questions.

In the second part of the segment Shezaadi voices her concerns about the ways in which Islamic women are portrayed on television. Initially she employs collaborative language demonstrating that she is in agreement with other on-line participants: 'yep you guys are correct', but then Shezaadi continues to give her personal experience with the specific programme: 'there was one particular scene …' Hence she expands the discussion by bringing in her own interpretation of the show. Shezaadi concludes her statement by using five exclamation marks thus conveying that she feels strongly about her particular point of view. In addition to using strong language and exclamation marks in order to index affect, these young women additionally include icons in their segments such as sad, happy or thoughtful faces. The next segment illustrates this:

(9) Aisha: (Female) Re: Where are the weapons of Mass Destruction?
Well we all know that they're just going after all the muslim countries and rulers one by one. We keep hearing about the 'roadmap' for the palestinians but thats never gonna happen … only muslim countries with WOMD are a threat to world peace …
thats why theyre now threatening pakistan and syria and completely ignoring India and Israel. And besides we know they're just gonna plant some weapons in Iraq anyway …
… This is there war against Islam after all because they start shaking every time a muslim stands up and says 'ALLAHUAKBAR'

In the Segment 9 above Aisha speaks against the narratives used to justify the war in Iraq. As Aisha expresses her point of view, she punctuates her statements with a thoughtful face ( ), then an unhappy face ( ), and concludes with a strong note 'because they start shaking every time a muslim stands up and says "ALLAHUAKBAR" (God is great) and a happy face ( ). Note also that Aisha uses capital letters which once again serves to display assertion.

These young women do not simply use these discussion threads to explore, argue and contest. The women often share their knowledge regarding various issues concerning Islam. The following excerpts show the ways in which they convey this.

Women as agents of knowledge

One of the ways in which the young women frequently exhibit their knowledge is by reference to varying websites and television and radio programmes:

(10) *Sara* (Female): Re: Hijab March on 17 June
Did anyone attend this march?
Hamza (Female): Re: Hijab March on 17 June
On the subject of hijab, a programme is coming on Channel 4 Called Islam Unveiled on Sunday at 12:10.
The presenter of the programme Samira Ahmed, said that the Qu'ran says that the women should cover their beauty but not wear hijab.
She says this towards the end of this audio clip.
http://www.bbc.co.uk/radio4/womanshour/12_01_04/wednesday/ram/item3.ra+m/C.

(11) *Shia sister* Re: Feminism
Assalam u alaikum
Islam gives more rights to women then west. I cannot list all of them here. I will give a link to an excellent book on women's right by a great scholar of Islam
http://www.al-islam.org/WomanRights/index.html

(12) *Maryam* (Female): Documentary Tonight
Assalam u Alaikum
Just thght of informing you guys about a documentary tonight, should be interesting, one not to be missed especially for non-muslims, and muslims of weak faith like myself. Channel 4, 6pm, Turning Muslims In Texas. It is only 30 mins long so try not to miss 15 mins of it.
Here's an insight into the programme:
http://www/Channel4.com/culture/microsites/B/believeitornot/texas1.html

These excerpts show that these young women are very well informed and knowledgeable about the various sites and spaces which carry relevant information. In doing so, they index self-empowering identities which challenge dominant discourses concerning women which often view Muslim women who live behind the veil as having no access to information. Moreover, the fact that these on-line participants accessed information outside formal classroom contexts further demonstrates their agency in their quest for knowledge and education.

In addition to providing information about various websites and links, these participants also display their knowledge by using Islamic vocabulary:

(15) *Khadijah* (Female): Re: Hijab enforcement
Assalam u Alaikum
Yeah I agree actually it can't really be forced because what happens is the girls take it off at school so what is the point In that. Islam does start at home with the right **tarbiyyah** and guidance. InshaAllah girls will wear hijjab out-side and in One should use **Hikmat**.

(16) *Maryam* (Female): Assalam u Alaikum Wr Wb
Surely an essential aspect of Islam is to enjoin the good and forbid the evil? If we as muslims fail to do this then **fithna** will prevail and will become the truth and truth will become falsehood.

(17) *Eazy D* (Female): But not all punishments are from the **hudud**. There are other punishments, which come from other principles of the Islamic penal code like **ta'zeer** and **jineeyat**. Like the punishment of those who do not pray is based on ta'zeer. Ta'zeer in essence is a type of punishment within the shar'ee which is left to the discretion of the **Khalifah**. So punishment of those who do not wear the Hijab in Public will fall into ta'zeer.

The use of Arabic words such as *'tarbiyyah'*, *'hikmat'*, *'fithna'*, *'Khalifah'* and *'jineeyat'* convey that these young women are knowledgeable about Islam at a deeper level than the merely cursory. In many ways the use of such Islamic terms in addition to the questions initiated by young women reveal their intention to explore Islam in serious ways. In other instances, young women display their knowledge by including quotations from various sources in order to justify their particular position. The following is an excerpt from a rather long segment. The participant voices her opinion against the recent ban on hijab:

(18) *Maryam*: Re: Chirac and the Hijab
Furthermore Jacques Chirac like many other ignorant people made the all too common mistake, that secularism was built on honouring women, but Islam was built on dishonouring women: **'A society's level of civilization is measured first and foremost by the position that women occupy in it', Chirac said last week.** (http://www.mpacuk.org/mpac/data/9c4bb9dc/9c4bb9dc.jsp)

> If the position of women does directly reflect her high status in society then only Islam is the most supreme and unequalled in the honor and dignity it showers upon them. I will briefly mention a few examples. Firstly women are honored with the law of the hijab:
> **Say to the believing men that they cast down their looks and guard their private parts; that is purer for them; surely Allah is Aware of what they do ... (Surah 24, 30–31)**

Thus Maryam includes several voices in this segment, Chirac's voice, then her own, and then God's. Thus, by including several sources to articulate her position, she shows that she not only has a strong position but that she is also able to counter some of the rhetoric by citing various sources in a convincing fashion since the next participant compliments Maryam, 'Well said', and the discussion comes to a close.

Conclusion

This chapter explored the discursive practices of young Islamic women during on-line discussion threads. By examining the linguistic practices of these women, the study shows that they convey their identities in a variety of ways such that they are assertive, knowledgeable, young, female and Islamic. In showing the varied types of identities that these young women invoke in interaction with other female as well as with male participants, the study aimed to debunk many of the stereotypes that position these women as subordinate and submissive. Furthermore, the study contributes to scholarly work on language and gender which does not always conceptualize women's speech and men's speech in dichotomous ways. Last, the study shows that women, who self-identify themselves as religious (Islamic) through language in this particular context, may have complex identities and need not be understood in monolithic ways.

Note

1 Located at www.TheRevival.co.uk

References

Bucholtz, M. (1999). Bad examples: Transgression and progress in language and gender studies. In: M. Bucholtz, A. C. Liang and L. Sutton (Eds), *Reinventing identities: The gendered self in discourse.* New York: Oxford University Press. (pp. 3–24).

Bullock, K. (2003). *Rethinking Muslim women and the veil: Challenging historical and modern stereotypes*. London: The International Institute of Islamic Thought.

Goodwin, M. H. (1990). *He-said-she-said: Talk as social organization among black children*. Bloomington: Indiana University Press.

Goodwin, M. H. (1999). Constructing opposition within girls' games. In: M. Bucholtz, A. C. Liang and L. Sutton (Eds), *Reinventing identities: The gendered self in discourse*. New York: Oxford University Press. (pp. 388–409).

Herring, S. C. (1993). Gender and democracy in computer-mediated communication. In: *Electronic Journal of Communication* 3(2).

Herring, S. C. (1996). *Computer-mediated communication: Linguistic, social and cross-cultural perspectives*. Amsterdam: John Benjamins.

Holmes, J., and M. Meyerhoff (2003) (Eds), *The handbook of language and gender*. Oxford: Blackwell.

Kamalkhani, Z. (1998). Reconstruction of Islamic knowledge and knowing: A case of Islamic practices among women in Iran. In: K. Ask and M. Tjomsland (Eds), *Women and Islamization: Contemporary dimensions of gender relations*. New York: Berg (pp. 177–93).

Khan, S. (2002). *Aversion and desire: Negotiating Muslim female identity in the diaspora*. Canada: Women's Press.

Mahmood, S. (2003). Ethical formation and politics of individual autonomy in contemporary Egypt. In: *Social Research* 70(3).

Mendoza-Denton, N. (1999). Turn-initial *no*: Collaborative opposition among Latina adolescents. In: M. Bucholtz, A. C. Liang and L. Sutton (Eds), *Reinventing identities: The gendered self in discourse*. New York: Oxford University Press (pp. 273–92).

Orellana, M. F. (1999). Good guys and 'bad' girls: Identity construction by Latina and Latino student writers. In: M. Bucholtz, A. C. Liang and L. Sutton (Eds), *Reinventing identities: The gendered self in discourse*. New York: Oxford University Press (pp. 64–81).

Panyametheekul, S., and Herring, S. (2003). Gender and turn allocation in a Thai *chat* room. *Journal of Computer-mediated Communication* 9(1). Retrieved 25 February 2004 from: http://www.ascusc.org/jcmc/vol19/issue1/panya_herring.html

Raudvere, C. (1998). Female dervishes in contemporary Istanbul: Between tradition and modernity. In: K. Ask and M. Tjomsland (Eds), *Women and Islamization: Contemporary dimensions of gender relations*. New York: Berg (pp. 125–43).

Werbner, P. (2002). *Imagined diasporas among Manchester Muslims: The public performance of Pakistani transnational identity politics*. Oxford: James Curry.

12

'Inshallah, today there will be work': Senegalese Women Entrepreneurs Constructing Identities through Language Use and Islamic Practice

Shartriya Collier

The linkages between religious and economic growth and development have been irrefutably established. Indeed, it is often asserted that the shift from Catholicism, and an accompanying worldview that condemns materialism, to Protestantism served as the foundation for the Industrial Revolution. Likewise, Islam is often viewed as a religion that supports trade and commercial exchange. This study is an ethnographic sojourn that uncovers how a group of Senegalese-American women entrepreneurs construct their identities through language use and Islamic practices. Set within the context of a hair-braiding shop in Philadelphia, this study utilizes in-depth interviews, participant observations, focus groups and other qualitative research data collection methodologies to illustrate the power relations and cultural balancing of identity that characterize this immigrant population.

Embedded within the Feminist post-structuralist theoretical framework, the study reveals how the hair-braiding shop serves as an enclave for the merging of two cultures, American and Senegalese. Moreover, this unique employment niche is examined as a means of gaining greater insight into a Senegalese female's identity transformation process as she journeys from one culture to another. More specifically, it reflects the linguistic negotiation of self that occurs through the participants' blending of English, Wolof, French and Arabic – based upon religious beliefs.

Upon entering the Astou hair-braiding salon, there is an oversized poster of a man dressed in striking white garb, who stares at the public through dark piercing eyes. His eyes are distant as if contemplating the hardships experienced by mankind. The man – Sheikh Amadou Bamba – is a political and spiritual figure in Sufi Islam. Bamba, the founder of Muridiyya Islam, is considered to be the most powerful Muslim to assume a stance against colonial rule in Senegal. Bamba was exiled from Senegal to Gabon by French officials for 33 years. It is the presence of this poster that first suggests that Islam remains a critical influence in the lives of these female entrepreneurs.

In modern-day Senegal, it is common practice to hang Amadou posters in local businesses. Most often hung next to doorways, this practice represents Bamba's role as a gateway to divinity. Although the African-American clientele who patronize the salon often ignore this subtle dedication to Islam; for the Senegalese-American workers and store owner, the poster is a constant reminder of religious devotion, struggle and power. The poster is symbolic of colonial oppression, success and freedom in a new, yet alienating country, the United States. Bamba's teachings espouse the virtues of pacifism and hard work – traits that are the very foundation of entrepreneurship. Although the entrepreneurial life provides more freedom of expression for many immigrant women, their lives are still fraught with difficulties for it is difficult to build and sustain a strong client base.

'Inshallah, today there will be work' encompasses the women's daily struggle to survive when the shop's clientele dwindles. In the midst of a discussion in Wolof, one is likely to hear the word 'Inshallah' – a phrase that means 'God's will' in Arabic. 'Inshallah today there will be work' serves as a plea for customers. In the absence of customers, Marie and Selah, the store owners, will receive no money. Thus the entrepreneurial effort of these women is buttressed through religious faith.

Marie and Selah are not outliers. Currently, there is a rapidly growing population of immigrant women who are using culturally based practices such as hair braiding as the basis for entrepreneurship. Yet these entrepreneurial efforts serve a role beyond the realm of the economic. Specifically, the ownership and operation of culture-based businesses serve as community networks that allow for the preservation of traditional Senegalese-based Islamic practices and cultural traditions. Additionally, within this enclave the Astou hair-braiding shop offers a religious and cultural community network for other Senegalese immigrants and speakers of Wolof, the most commonly used language in Senegal. These networks provide a perfect context that can be used by Feminist

Postructuralists in order to gain a greater understanding of the way that bilingual, religion-oriented women negotiate new 'discourses' as English language learners.

Although immigrant women often experience conflict as they journey between their two worlds – the mother culture and the host culture – the African women presented in this study are able to exert power and control over their daily linguistic, cultural and religious practices. However, perceptions of what it means to be Muslim and practise Islam vary between the older participants and the younger participants. More concretely, it appears that the facility with which the women adopt the English language is directly linked with the strength of their commitment to Islam. In particular, the use of English by the younger participants reflects a partial transition to an American identity, and with it, a shift towards a less conventional perspective of Islam. In the following sections, the implications that the study of women entrepreneurs and religious identity has for Postructuralist theory is highlighted. Additionally, an analysis of the manner in which the participants construct their religious identities and the role that the use of English plays in the process is described. Last, the implications of this study for researchers and educators in Gender Studies, Language Education and Religious Studies are outlined.

Feminist post-structuralism: power, agency and religious freedom

Communities of practice, whether in the workplace, church or home require that speakers understand the norms of interaction that will allow them to participate as members of a specific community (Holmes and Meyerhoff, 1999). While women in workplace discourses have been studied extensively (Tannen, 1994; Norton Pierce, 2000; Goldstein, 2001; Holmes and Meyerhoff, 2003), most findings reveal that immigrant women in the workplace are viewed as triply disadvantaged. That is, they are stigmatized by their gender; they are viewed as incompetent because they are second-language learners; and, in many cases, these women are considered as inferior because they are racial minorities.

Feminist Postructuralism focuses upon power and agency. It therefore offers an illuminatory framework that can be used to further conceptualize how identities are constructed within the hair-braiding salon. Language as a social practice means that language is a place 'where our sense of ourselves, our subjectivity is constructed' (Weedon, 1987, p. 21). The discursive practices in workplace settings represent the power relations between people of diverse races, ethnicities and genders. Postructuralists view language from an interesting perspective. This framework relies upon the conclusions of Bourdieu (1991) who argues that language is a form of symbolic capital which can be converted into economic capital. In institutional settings where women are working in subordinate positions, they are forced to succumb to many of the American cultural practices that have been dictated by society at large. For example, the management of workers occurs in English. Workers are also forced to conform to American workplace dress codes, and American cultural perceptions are used to define proper conduct. Such an environment does not, of course, allow many Muslims the time required for prayer. In contrast, business ownership provides immigrant women with the power to control their cultural practices in clothing, language use, and religious behaviour. For the Senegalese women presented in this study, the power to construct social meaning rather than have social meaning constructed for them manifests itself in when they choose to pray, the right to wear traditional Senegalese clothing, and the power to hang posters of religious figures upon the walls of their workplace.

Weedon (1987) argues that Feminist Postructuralism must 'pay full attention to the social and institutional context of textuality in order to address power relations in everyday life' (p. 25). Thus, within entrepreneurial settings such as the hair-braiding salon of the Senegalese-American women used in this study, power relations were manifested in their everyday interaction with predominantly English-speaking customers and with their Wolof/French/Arabic-influenced workers. While the use of Wolof, French and Arabic may not be viewed as linguistic capital by these customers, competence in these languages for the central participants in this study was directly correlated with maintaining ties to their Islamic religious practices and involvement in the community of other Muslims. The qualitative data described below reveals the role that language played in the Islamic practices of the subjects. The discursive spaces of the three women – Marie, Selah and Fatima – also illuminate the contrasting hegemonic practices of the older and younger participants. However, before examining the data it is necessary to briefly review how language use is dictated by colonialism and religious practice.

Senegalese women, Islamic practice and language use

Sufi Islam contrasts greatly with the Islam practised in the Middle East. Most Senegalese are associated with three Sufi brotherhoods: Tijaniyya, the Mourides and Qadiriyya. The founder of each brotherhood is believed to possess extraordinary spiritual powers also known as 'baraka'. However, Islamic practice in Senegal is integrally linked to the use of Wolof and Arabic. Most importantly, 'Wolof nationalism, African traditional religion and Sufi Islam have all been combined together in the hothouse of French colonialism to produce the unique folk Islam of Senegal' (Jackson, 2004). Thus, with regard to the African women in this study, it was necessary to determine whether the impact of colonialism created attitudes of resistance towards speaking English. Additionally, the study sought to determine how the women in this case study negotiate their religious practice in order to maintain connections to Senegal.

Literature on African history reveals that Africans in general appear to embody contradictory attitudes regarding the learning and use of English. This is not surprising given that traditions regarding language policy in Africa are highly complex. Thus colonialism has left many Africans with a resistant attitude towards using English, French and/or Portuguese since these were the dominant colonial languages in Africa. Alexander (1996), in his discussion of English use in South Africa, argues that English is associated with 'all of the baggage of colonialism and apartheid'. In response to this association, there have been multiple movements to advocate for a native language as the lingua franca for the entire continent. The underlying notion for this 'native tongue' advocacy is that it will help to unify the people. Nonetheless, Africa has more than 1200 spoken languages or dialects. Senegal alone uses French and Wolof as the lingua franca. However, there are at least three other languages spoken, Siegel (1999, p. 2) best describes the dichotomy of language acquisition and use in her discussion of English teaching in Senegal by saying: 'Imagine an infant in Dakar, with Serer or Fulani speaking parents, sisters and brothers who speak Wolof, their parents' language, and neighbors who speak Wolof and listen to French on the radio and television.' Thus, for the African women who are the subjects of the current study, even before immigrating to North America the acquisition and use of English had political, social and economic stigma attached to it.

Methodology

This qualitative study took place over a period of three years. The primary source of data collection was six hours of weekly observation at the Astou African hair-braiding shop. In addition to using extensive field notes, the main participant of the study was shadowed outside of the hair-shop setting. Thus interactions that occurred on the street, at the market or at home were analysed to gain greater insight into the individual's use of language and her religious and cultural identity. Second, the subjects for the study were interviewed in order to gain detailed information regarding the circumstances of their migration to the United States and the cultural stressors that may have impacted their learning of English and their religious identity. Two formal interviews and many informal interviews allowed the personal voice of the participants to be fully examined. The non-English-speaking women were interviewed using a translator. The interviews were later transcribed. Finally, observation was used to identify changes in English fluency among the women and to identify efforts to maintain Senegalese culture and Islamic practices.

The interviews and observations were reviewed and coded in order to analyse the reoccurring trends that arose in the study. Moreover, a number of approaches were used in order to further triangulate the outlined research questions. As mentioned, targeted secondary data was collected and content analysed in order to uncover background information about Senegalese migrants in general. This data was collected from the US Immigration Service and from the Senegalese Embassy located at 2112 Wyoming Avenue, NW in Washington, DC, 20008, (202) 234-0540. Other research tools that were used were an analysis of an African immigration exhibit at the Balch Institute and a symposium on African Women at Temple University. Also the participants interacted in a focus group with other Senegalese women aged 20–27. This provided insight into diverse perceptions of cultural identity amongst these women.

Participants

The central participant in this study was Marie. Marie is a 22-year-old Senegalese immigrant. Marie had been in the United States for three years at the time of the study. She has lived in various cities throughout the United States including New York, Philadelphia in Pennsylvania and Baltimore in Maryland. Of three siblings, she is the only one who followed her mother to this country. Her older sister remained in Senegal to take care of her home and the rest of the family. Last year, Marie married a man from Senegal. She does not have any children. Prior to coming to the United States, Marie studied English in high school. Marie studied in college for one year.

Another participant in this study was Marie's mother, whom I refer to as Selah. Selah was 47 years old at the time of the study. She came to America five years ago to be with her husband, who left Senegal a year earlier. Marie's mother speaks very little English. She also travels frequently to Baltimore and New York City. Selah owns the hair-braiding shop, but primarily allows her daughter to run it.

Last, Marie's Aunt Fatima is an additional participant in the study. Fatima is an extremely tall woman in stature. She had been in this country for three months at the time of the study. Fatima began braiding hair professionally upon her arrival to the United States. Fatima speaks Wolof and French.

Data analysis

Astou hair-braiding shop: a community of practice

Based upon various observations of the shop, a focus group interview, and individual interviews with the central participant, Marie, a number of observations can be made regarding the degree to which the business served as a community of practice. (Unfortunately, Wolof translations and transcriptions were not available. Therefore, most data involving Selah and Fatima are based upon observations.) The observational data were analysed using Content Analysis and Ethnography of Speaking based upon Dell Hymes (1972). This method is useful for gaining greater insight into the community environment and the norms of interaction in the salon. Interviews were analysed using Critical Discourse Analysis according to Fairclough (1995). Critical discourse analysis allows for the examination of power within this context. A brief account of the norms of interaction in the shop has been provided. Marie's use of English is then examined as it correlates to her identity and perceptions of religion. Finally, various observations of Selah and Fatima that demonstrate how they practise religion within the context of the shop and how this correlates with their use of English are described. The following is a brief description of the salon:

> A small white couch is angled against the wall facing the television. There are three customer chairs with leather stools placed directly behind them. On an average day, Marie will have three to four customers, primarily African American, who solicit a variety of hair-braiding styles. The common language of business exchange with customers is English. Nonetheless, one may often hear the sounds of traditional Senegalese music or tapes of music imported from Dakar. The women, Marie, Fatima, and transient family members, discuss life in their native language, Wolof. Upon entering, the store family members greet each other in French and/or Wolof.
>
> (Field note, 3 October 2001)

Marie: maintaining an African identity through minimal use of English

Marie, as the most dominant English speaker, has a different customer interaction from her mother and Fatima. To explore this it is necessary to look at Marie's daily interaction with customers. The following are three different excerpts from field notes:

Customer: How much for box braids?
Marie ((Stands up and walks to the display)): What size? We have this one, this ...
Customer: That size is good.
Marie: 150
Customer: Are you open on Fridays?
Marie: Yes. ((responds walking back to her stool))
Customer: Can I come in at 10:00?
Marie: That's fine.
Customer: How long will it take?
Marie: About five hours.

The phone rings. Marie reaches to her side to pick up the phone. 'Yes, What's wrong? It's coming out? No, don't take it out. You come here; I'll fix it for you. No, o.k. Then Thursday.'

(Field note, 1 November 2002)

((A woman with a blue-jean dress and long hair walks into the store))
Marie: Hello
Customer 2: How much for cornrows straight back?
Marie: With extensions?
Customer: Yes
Marie: 85$ (The girl with the long hair turns to walk) take a Card.

All three of these data demonstrate that Marie's interaction with her customers is limited solely to style and price negotiation. In each segment, the customer initiates the interaction with an interrogative such as 'How much' or 'Can I'. Marie's utterances are limited to two and three phrases that are solely responses. (Except in the last segment where Marie tells the customer to 'take a card'.) Even in the telephone call, there is no introduction such as 'Hello, how are you'. Thus, Marie's access to mutual engagement is very limited (Wenger, 1998). Yet, even though Marie is not mutually engaged through shared history or relationships that span beyond the business arena, she is experiencing a trade-off of engagement that allows her to negotiate power and define herself as the shop's owner and lead stylist. For example, in the excerpt above Marie's use of commands such as 'take a card' and 'no, don't take it out' reveal that she must position herself as an expert in her field who has knowledge of what the customer should and should not do.

As this section reveals, although Marie has limited opportunity to practice English, it demonstrates her self-positioning. By positioning herself as an entrepreneur, Marie simultaneously establishes an American identity. The concept of imagination as defined by Wenger is 'the ability to expand ourselves by transcending our time and space and creating new images of the world and ourselves' (1998, p. 176). Thus imagination is exemplified through these exchanges. These abridged interactions assume additional meaning when compared with conversations that take place in Wolof and French:

> The phone rings. 'Ca va', Marie answers. 'Jirry gif'. Her eyes widen and she raises her voice. Her voice is now growing louder and the intonations in her speaking are flowing up and down. She begins to move her fingers and shoulders as if receiving exciting news. She hangs up the phone. 'That was my sister.'
>
> (Field note, 30 November 2002)

Marie answers the phone using the French greeting, and the Wolof phrase 'jirry gif' for thank you. Marie's expression and use of Wolof and French serves as a bond to her home culture. Nonetheless, the loneliness that distance creates cannot be defined by any one action. Yet, it can be expressed through a language that one relates to childhood and to home. Marie's language, Wolof, connects her to her past in Senegal and to her present life. One evening while eating dinner, Marie was referring to a story that she learned in her childhood. Yet, she could not find the words to properly translate the story:

> Marie looks at her husband and speaks in Wolof: 'You know after the lion takes that boy to the village, there is a lesson to the story, but I don't know how to explain it in English.' Marie's husband looks up, 'the boy had to learn to live with himself before he could learn to love someone, something like that.'
>
> (Field note, 12 December 2002)

Marie had trouble fully translating concepts outside of the business arena. Therefore, she is unable to translate many of her childhood experiences.

Thus, Wolof becomes the link that is a focal tool for her Senegalese life. Hoffman (1989) explains the void that English acquisition can create, as a gap not fully closed. 'But I begin to trust English to speak to my childhood self as well, to say what has so long been hidden, to touch the tenderest spots' (p. 274). Possibly, Marie's journey of self-transformation will eventually lead her to negotiate meaning as she begins to invest in becoming an English speaker. Without her English, she would not have a successful business.

The question that arises is how can Marie extend her English-language practice in the hair salon? As the only English speaker in the shop, Marie is forced to interact with her customers and negotiate prices for the other stylists. This analysis suggests that Marie serves as a cultural broker. Most importantly, this trade is a cultural tradition that has been passed down from mother to daughter for centuries. The current immigrant studies on power reflect (McGroaty, 1996; Mckay and Wong, 1996; Heller and Martin-Jones, 2001) the subjects' standpoint as 'participating from the peripheral'. Marie, her own boss, is able to position herself from a powerful vantage point.

Selah and Fatima: when religion intersects with tradition

Selah and Fatima's experiences vary due to their limited English. This conclusion is revealed in the following, an interview between Selah and the researcher:

R: So who is that man?
S: She great leader, um, she fight
R: He
S: He go to Gabon to fight for, mum (trying to find the words)
To fight for the people he know want to be (uses hands gestures like an explosion) they kill him. He went to Gabon.
R: So he was a leader for the people against the French, he's dead now?
S: They kill him. But every year we people come, all de Muslims people dey come from all de cities to Dakar.

In reviewing Selah's struggle to convey the story in English, it is important to focus upon the use of the pronouns he, we and they. While it is clear the 'he' is referring to Bamba and 'they' refers to the French, the use of 'we' allows the participant to position herself as a Muslim. Additionally, Selah's efforts to use English reveal how the colonial conflict appeared to the Wolof people of Senegal. Pennycook (2001) provides additional understanding of this response in his discussion of resistance as a response to power in the postcolonial area. The author defines colonialism as not simply economic or political but also as 'a state of mind' (Pennycook, 2001, p. 67). Inevitably, when discussing African immigrants and language learning, the impact of colonialism must be considered. Rampton (1995) discusses the use of language as a tool of oppression. The concept of 'belonging' often places the speakers of English as a second language with a group to which he or she may not necessarily 'belong'. For example, many African immigrants are assumed to be African American and learn African-American Vernacular English. Selah's sense of belonging in the United States is largely connected to having her own business and remaining a Muslim. The following is a telephone conversation in English that took place between Selah and a Sudanese man:

> *S*: Yes the women, she is Moroc, Moroc, Moroccan. She's a good Muslim woman. Do you have a house? Good credit? I tell her that ... no just good Muslim man. OK. Asalaamilakum

After the conversation took place Marie informed me that the Sudanese man wanted Selah to find him a wife. The connecting factor between Selah, the man from the Sudan and the Moroccan woman is their devotion to Islam. Selah's inquiry into the man's financial matters demonstrates that she is conducting a negotiation although not necessarily economic. The man's response to her inquiry about his financial success is that it is irrelevant. She reaffirms this by a brief 'OK' and a farewell in Arabic, the language of Islam.

In contrast, the following are excerpts of Selah interacting with African-American, non-Muslim customers:

> An African-American woman wearing a black coat walks in the shop. She looks Selah in the face, 'How much for cornrows going back with a part in the middle?' Selah looks over and points to Marie. 'No English, ask her', she responds.
>
> (Field note, 17 November 2002)

> Selah is sitting in the door closest to the entrance. An older woman with a long black wig walks in with her child. She looks at Selah, 'I wanted to get my daughter's hair braided with no extensions.' Selah responds, 'You pay 60$'. 'But she's not getting extensions, it's her

> own hair', the woman answers. Selah responds, 'You wait for my daughter'. The woman sits down. Marie returns to the room.
>
> (Field note, 6 February 2002)

> Marie is working on a customer's hair. A new customer, a young black male, walks into the room. 'How much for cornrows?' '50$', Marie responds. When can I get it done? 'Now', Marie answers. She speaks something in Wolof to Selah. Selah waves the male over to take a seat.
>
> (Field note, 17 February 2002)

In the first excerpt, Selah quickly refers the customer to Marie. However, in the second excerpt, she attempts to use English by quickly establishing a price. When the customer asks for further clarification in both interactions, Selah refers the client to Marie for negotiation. However, when Marie is questioned about this, it is discovered that her mother is 'technically' the owner of the shop. However, as witnessed in the excerpt above, Selah refers all customers to Marie.

Generation discontinuities in English acquisition

As previously mentioned, Senegalese women immigrants arrive in this country already fluent in two or more languages. With Wolof as their first language and French as the official language of Senegal, Senegalese women have personal knowledge of the dynamics and varied contexts of language, speech and identity. Many of the hair braiders who arrive in this country have already had exposure to the English language. This section will explore the varied use of language between Marie, Selah and Fatima. For Marie, English was used out of necessity, but for Selah and Fatima, English was used out of choice. For example, the following excerpt occurred between Marie and Fatima:

> The phone rings 'Ca va', Marie answers. 'Hold on', Marie holds the out to Fatima. 'Fatima, telephone', Marie calls. Fatima stands up smiling. 'Hello, Hello ... What?', Fatima gets on the phone and sits down. She then switches back to Wolof.
>
> (Field note, 1 October 2001)

This was the first time Fatima had been observed using English. This field note is dated 1 October 2001. When this speech act occurred, observations had been taking place for a period of six months. This was a similar situation with Marie's mother Selah. Throughout the duration of the study, Selah only spoke English randomly. However, these occasions revealed that Selah had a greater knowledge of English than she revealed:

> 'Did your mother used to work on 125th Street in Harlem?', I ask Marie. Before Marie can answer, her mother shakes her head, 'yes, yes', she answers. 'Does your mother still live in Philadelphia?', I ask looking at both Marie and Selah. 'Yes', Marie's mother again begins answering.

Although on both occasions, Selah and Fatima answered with a simple 'yes' or 'no', this dialogue indicates that there is some level of English comprehension on behalf of both of the women. In Fatima's case, she may have understood the meaning according to context. However, Selah demonstrated through body movement and responses that she understood what was occurring. These behaviours can also be viewed as correlating with religious devotion on behalf of the older women. All three women are in fact Muslim. However, Marie was never witnessed actively partaking in the act of prayer or Salat. The following data reveal displays of religion for Fatima and Selah:

> Fatima takes the mat in her hands. She fixes the scarf wrapped around her neck so that it completely covers her hair. She says something in Wolof to Marie. 'Should I leave?', I ask noticing that she's about to pray. 'No, she was just asking which way was east', Marie responds.
>
> (Field note, 6 February 2002)

> Marie's mother comes out of the bathroom with a straw mat and Fatima's scarf wrapped around her shoulders. She covers her hair, I see her stand up, kneel, and put her head to the floor. She stops and sits quietly for a moment.
>
> (Field note, 18 January 2002)

In contrast, this is an excerpt taken from an interview with Marie regarding religion:

R: Marie, I noticed sometimes when I'm here your mom and your aunt are praying. Do you do that too?
M: Yes, in the morning, but not five times a day. What about you, what's your religion?
R: Well, I was raised a Christian but I'm non-denominational.
M: Where I'm from, we believe in Allah and the prophet Mohammed.

R: How come you don't cover your hair?
M: Not everybody does that in my country.
R: But your mother prays five times everyday.
M: But she's very religious. The older women do that more.

This interview is revealing in several ways. Marie, as related to her religion, is devoted. However, she may not be as traditional as her mother. In some respects, her relaxed attitude toward the religion is symbolic of her positioning as an American woman and as a Senegalese immigrant. Marie's English also allows her to interact with fellow American Muslims. This opportunity broadens her access to the language. For her mother and her aunt, English has an impact on their religious devotion between non-Wolof speaking Muslims. On the other hand, code switching between Arabic and English commonly occurs. Fatima and Selah do not have the same motivation to learn because it is not a necessity for their survival. Although both women live in the United States, they have Marie and her younger, more fluent family members to speak for them. Yet members of their own linguistic group most commonly surround them. Thus Marie is the nucleus of the Astou hair-braiding shop. Simultaneously, she has the task of mediating her identity as a young Senegalese woman in America. Clearly, community for these participants is deeply embedded in their sense of community as Senegalese women.

The following is an excerpt from a focus group with Marie, her cousin Fatou and Danjou. This demonstrates the way the younger women positioned themselves as Americans in contrast to Selah and Fatima.

Marie: Yes, they stick more to the way we do things in Senegal, but the young girls, we like to look pretty and dress anyway.
Marie: I miss my family and friends in Senegal, but I like shopping and all the things we don't have in Senegal.
Fatou: Yes, some things we can't do in Senegal, but we can do here.
Marie: You can make more money here in a week than you can in two weeks in Dakar.

Community for Fatou and Marie is using all of the advantages they did not have in Senegal, but have in the United States. Marie supports this statement in her quote above, 'You can make more money here in a week than you can in two weeks in Dakar.' Being Senegalese represents community for the women who were involved in this focus group. This level of acculturation is represented by the women's interest in 'shopping and making money'. Marie confirms this statement again by stating, 'Yes, they stick more to the way we do things in Senegal, but the young girls, we like to look pretty and dress up.' However, looking pretty and dressing up violates traditional Islamic values for a female.

When relating to Senegal, her main complaint involves missing her family. Yet her preference for material prosperity allows her to function as an American in the United States. Marie also relates this American/ Senegalese identity to traditional versus non-traditional dress and religious practices. Marie finds that she has the personal power as an American to choose how she practises. The imported cultural trade of hair braiding allows her to use a traditional trade in a manner which is beneficial to her financial success in this country. It is clear that Marie has a strong sense of control over her own life. Control is one of the reasons why Marie prefers to work for herself rather than someone else.

Linguistic resistance

Why don't Selah and Fatima choose to speak English? Morrow (1997), Kouritzin (2000) and Wong (1991) would conclude that these women have not been provided with an opportunity for equal access. Moreover, they found that older immigrant groups show resistance to learning English because if they give up their language, they perceive themselves, in many ways, as giving up their culture. These feelings are manifested as attitudes of ambivalence. Fatima and Selah's attitudes of ambivalence impact their English acquisition. Thus Marie is relied upon as a translator. This trend is revealed in the interactions described below:

> A woman walks into the room, holding a paper bag. 'Do you want any products today?', she asks looking at Marie. Marie translates this to her mother. Marie's mother looks from where she is sitting on the couch to the bag; she hesitates and says something in Wolof. Marie looks at the woman, 'Come back tomorrow. Today it's slow'. The woman answers, 'Well I won't be able to come back until next month.' Marie translates this to her mother. Her mother moves her eyes around as if contemplating. 'Come back', she says in a thick accent.

Once again, this excerpt reveals that Marie's mother has the ability to understand English. However, she uses Marie as a translator. Swigart (2000) explains this occurrence by arguing that younger generations have the ability to adapt more quickly to a new society. The younger generations are often called upon to help parents understand American culture and language. Considering this explanation, one must reflect on language motivation. Spolsky (2000) would argue that Selah and Fatima's linguistic resistance reflects a fear of *anomie*, or succumbing to an American identity. Do Fatima and Selah feel threatened or simply uninterested? Kerswill (1996) verifies that older adult groups face more difficulties when acquiring a new language. Tabouret-Keller (1997) also addresses parental dependence on children for linguistic survival as a multilingual/multicultural person. However, negotiation (Pavlenko, 2001) of her identity will, hopefully, be a painless process that allows Marie to determine her own identity. Marie's attainment of multiple registers (Koven, 1998) provides identity alternatives for Marie within the context of the shop as well as outside of the shop.

Discussion

This study tells the story of the complex transformation that Senegalese American Muslim immigrant women undergo not only while working in hair-braiding shops, but also in acculturating into American culture in general. Yet, it appears that one price that Marie pays is a reduction in her adherence to the tenets of Islam. Nevertheless, the hair-braiding shop is a community of practice whose success is contingent upon Marie's investment in the English language. These circumstances do indeed place her in a position of power. Although researchers such as Koven (1998), Davies and Harré (1990) and Tabouret-Keller (1997) identify and explore the dynamics of identity from a transformative position, they do not view them from the perspective that power positions can and are already established by certain immigrant groups. Furthermore, cultural maintenance occurs on a daily basis through language, religion, trade and customer relations. This study demonstrates that the women have a very strong sense of their Senegalese identities. Yet, depending on generational patterns, they may negotiate a more pronounced American identity. This correlates to the findings of Spolsky (2000), Kouritizin (2000) and McKay and Wong (1996) whose theories of ambivalence, motivation and access are applied to explain the resistance of the older women to English acquisition.

Implications

This study also has implications for both the fields of Second Language Acquisition and Religious Studies. It reaffirms the notion that older immigrant groups are often resistant not only to language learning, but to the host cultures to which they have migrated. This research also examines how language learning in the context of the working environment and religious identity are fluid due to engagement, alignment and negotiation. Furthermore, as Marie's linguistic registers evolve, her identity will continue to transform.

The hair shop as an enclave for family meetings, worship and extended family is crucial to these Senegalese women. It provides a safe haven for generating income and networking with other women. Life in the shop is a challenge due to low customer interaction. Nonetheless, the women realize that by working together, they can be more successful. The vivid story that remains constant in this research is that language and religion in practice are dependant upon a plethora of social, cultural and even generational factors. Marie is striving to engage in linguistic and cultural negotiations in her shop. Through these efforts, she can redefine her position in this society. For Fatima and Selah, their devotion to their Senegalese past, present and future are consistent patterns that maintain their Senegalese identity but place boundaries on their American transformation process.

The phrase 'Inshallah, there will be work' draws attention to the long-ignored experience of a growing American cultural group. The unique trade and cultural practices that occur within African hair-braiding shops may serve as a maintenance model for future immigrant groups. Marie's linguistic and diverse cultural identity allowed her to successfully navigate in two worlds. As educators, we must seek ways to further assist immigrants to acquire English for the workplace and beyond. Additional questions that may be initiated from this exploration are:

- What are ways that we can motivate older generations of immigrant groups to expand their English repertoires?
- How can we create entrepreneurial opportunities for the other immigrant groups who are rapidly populating this country?
- Can customer relations between immigrants and non-immigrant groups be improved?
- Are English language studies within the social context of work useful for future teachers?

The observational study described in this chapter is designed to serve as a starting point for further inquiry into these issues.

References

Alexander, N. (1996). LANGTAG and UNISA's language policy. Panel input prepared for the conference Towards a Language Policy for Unisa, 23 February, Pretoria.

Bourdieu, P. (1991). *Language and symbolic power*. Cambridge: Polity Press.

Bureau of the Census. (2000). *Highest-ranking countries of birth of US foreign-born populations*. United States.

Carliner, G. (2000). The language ability of US immigrants: Assimilation and cohort effects. *International Migration Review* 34(1), 158–82.

Davies, B., Harre, R. (1990). Positioning: The discursive production of selves. *Journal for the Theory of Social Behavior* 20, 43–63.

Dodoo, F. (1997). Assimilation difference among Africans in America. *Social Forces* 76(2), 527–46.

Donaldson, L. E., and Kwok, P. (2002). *Post-colonialism, feminism, and religious discourse*. New York: Routledge.

Fairclough, N. (1995). *Critical discourse analysis: The critical study of language*. London: Longman.

Goldstein, T. (2001). Researching women's language practices in multilingual workplaces. In: A. Pavlenko et al. (Eds). *Multilingualism, Second Language Learning, and Gender*. Germany: Mouton De Gruyter.

Heller, M., and Martin-Jones M. (Eds). (2001). *Voices of authority: Education and linguistic difference*. Westport, CT: Ablex.

Henry, A. (1988). *Taking back control: African Canadian women teachers' lives and practice*. New York: New York Press.

Hoffman, E. (1989). *Lost in translation: A life in a new language*. New York: Dutton.

Holmes, J., and Meyerhoff, M. (Eds). (2003). *The handbook of language and gender*. Oxford: Blackwell.

Holmes, J., and Stubbe, M. (2003). 'Feminine' workplaces: Stereotype and reality. In: J. Holmes and M. Meyerhoff (Eds), *Handbook of language and gender*. Oxford: Blackwell (pp. 573–99).

Hymes, D. (1972). *Foundations in sociolinguistics: An ethnographic approach*. Philadelphia: University of Pennsylvania Press.

Jackson, M. (2004). Islam in Senegal. The Mouride Brotherhood. Retrieved from http://www.bcconline.org/wolof/Mouridism%20article.htm

Joy, M., and Dargyay, E. K. (1995). *Gender, genre and religion: Feminist reflections*. Ontario: Wilfred Laurier University Press.

Kainola, M. A. (1982). *Making changes: Employment orientation for immigrant women*. Canada: Cross-Cultural Communication Centre.

Kerswill, P. (1996). Children, adolescents, and language change. *Language Variation and Change* 8(2), 177–202.

Kouritzin, S. (2000). Immigrant mothers redefine access to ESL classes. *Journal of Multilingual Contradiction, Ambivalence and Multicultural Development* 21(1), 14–32.

Koven, M. (1998). Two languages in the self/the self in two languages: French-Portuguese bilinguals' verbal enactments and experiences of self in narrative discourse. *Ethos* 26(4), 410–55.

McKay, S., and Wong, F. (1996). Multiple discourses, multiple Identities: Investment and agency in second language learning among Chinese adolescent immigrant students. *Harvard Educational Review* 3, 577–608.

McGroaty, M. (1996). Language attitudes, motivation, and standards. In: S. L. Mckay and N. H. Hornberger (Eds), *Sociolinguistics and language teaching*. Cambridge: Cambridge University Press.

Morrow, N. (1997). Language and identity: Women's autobiographies of the American immigrant experience. *Language and Communication* 17(3), 177–85.

NorKunas, M. (1999). Women, work, and ethnic identity: Personal narratives and the ethnic enclave in the textile city of Lowell, Massachusetts. *Journal of Ethnic Studies* 15(3), 145–78.

Norton Pierce, B. (1995). *Social identity, investment, and language learning. TESOL Quarterly* 29(1), 9–31.

Norton Pierce, B. (2000). *Identity and language learning: Gender, ethnicity, and educational change.* Singapore: Pearson ED.

Ochs, E. (1992). Indexing gender. In: A. Duranti and C. Goodwin (Eds). *Rethinking context* Cambridge: Cambridge University Press (pp. 335–58).

Pavlenko, A. (2001). Bilingualism, gender and ideology. *The International Journal of Bilingualism* 5(2), 1176–51.

Pennycook, A. (2001). *Critical applied linguistics. A critical introduction.* Mahwah, NJ: Lawrence Erlbaum.

Rampton, B. (1995). *Crossing: Language and ethnicity among adolescents.* London: Sage.

Siegel, M. (1999). Some observations of education and English teaching in Senegal. *TESOL Matters* 9(2) Retrieved 7, Dec. 2002 from http://www.tesol.org/ pubs/articles/1999

Spolsky, B. (2000). Language motivation revisited. *Applied Linguistics* 21(2), 157–69.

Swigart, L. (2002). *Extending lives: The African immigrant experience in Philadelphia.* Philadelphia, PA: The Balch Institute for Ethnic Studies.

Tabouret-Keller, A. (1997). Language and identity. In: F. Coulmans (Ed.), *Handbook of sociolinguistics.* Oxford: Blackwell (pp. 315–26).

Tannen, D. (1994). *Talking from 9 to 5. Women and men in the workplace: Language, sex, and power.* New York: William Morrow.

United Nations. (1999). *Estimated illiteracy rates in selected countries.*

US Immigration and Naturalization Service. (2000). *Statistical yearbook.*

Wagner, Stephen J. (1981). America's non-English heritage. *Society* 1(2), 13–35.

Weedon, C. (1987). *Feminist practice and postrucuralist theory.* Oxford: Blackwell.

Wenger, E. (1998). *Communities of practice: Learning meaning and identity.* Cambridge: University Press.

Wong, F. (1991). When learning a second language means losing your first. *Early Childhood Research Quarterly* 6, 323–46.

13

Gender, Hebrew Language Acquisition and Religious Values in Jewish High Schools in North America*

Debra Cohen and Nancy Berkowitz

The present study examines gender differences in Hebrew language performance, and attitudes towards learning Hebrew in different groups within the North American Jewish community. The chosen groups differ in their religious affiliation (non-Orthodox vs. Orthodox) and gender class composition (co-ed vs. segregated). These groups range from the non-Orthodox groups, which see gender equality as a central modern-Jewish value (Fishman, 2000), to certain Orthodox groups which see gender inequality as a Jewish traditional value that should be preserved. Segregated education usually indicates agreement with the latter. We found a significantly wider gap in language performance, learning goals and attitudes between boys and girls, in favour of the girls, in the Orthodox/segregated group than in the other two groups (non-Orthodox and Orthodox/co-ed). This finding supports the poststructuralist approaches which view gender as a social, historical and cultural construct that mediates between culture and language behaviour (Gal, 1991; Piller and Pavlenko, 2001). We also showed that learning goals of the learner play a central role as mediators between gender and language behaviour.

* The preparation and publication of this article was made possible by a grant from The Memorial Foundation for Jewish Culture. We also wish to thank the AVI CHAI Foundation, Hebrew College and the NETA project for giving us permission to collect information on the process of Hebrew learning from the NETA curriculum students. We wish to thank the school principals, NETA coordinators at the schools, teachers and students for giving from their time and thought for this important initiative.

Theoretical background

One agreed fact is that in second language college courses female students outnumber males at all levels, with the numeric gap widening at the more advanced levels (Chavez, 2001). Other research projects show that girls in second language courses are higher achievers, more motivated, have a more positive attitude, and report greater overall strategy use than boys (Gardner and Lambert, 1972; Burstall, 1975; Politzer, 1983; Boyle, 1987; Oxford and Nyikos, 1988; Spolsky, 1989; Ellis, 1994; Huebner, 1995; Baker and MacIntyre, 2003). Critics claim that the superiority of women in the area of second language acquisition reflects only the gap in performance on written tests, whereas in oral tests the opposite is true (Chastain, 1970; Boyle, 1987; Brecht and Ginsberg, 1993). Others found that women also outperform men on oral tests (Nyikos, 1990; Shiue, 2003), or found no differences between the genders (Bacon, 1992). The search for a deeper understanding of variables that affect achievement brought upon a shift in the area of second language learning and general learning, from focus on achievement to focus on motivation, learning goals and self-efficacy as mediating variables between gender and achievement. These variables depend highly on culture (Pajares and Giovanni, 2001). In the same vein, recent poststructuralist approaches view gender as a social, historical and cultural construct, which mediates between culture and language behaviour (Gal, 1991; Piller and Pavlenko, 2001). Here the locus of study shifted to ideologies of language and gender across cultures as well as over time within a culture (Eckert and McConnell-Ginet, 1992; Bonvillain, 1997; Holmes and Meyerhoff, 1999; Piller and Pavlenko, 2001). To date, most of the research that was aimed at investigating the cultural hypothesis of language acquisition focused on adults in separate immigrant societies. Research showed that within immigrant communities, gender differences in use of heritage language and English are strongly connected to the cultural context (Sole', 1978; Klee, 1987; Zentella, 1997; Lynch, 2004).

In this study we looked into possible gender differences in achievement, motivation, self-efficacy and general satisfaction with the Hebrew language course. The motivation of the student was tested according to achievement goal theory, which focuses on the distinction between mastery versus performance goals, broadly defined as the attempt to develop, improve or acquire skills, proficiencies and understandings versus the strivings to maintain self-worth through the demonstration of ability and avoidance of incompetence (Dweck and Elliott, 1983; Spence and Helmreich, 1983; Nicholls, 1984; Butler, 1987; Dweck and Leggett, 1988; Maehr, 1989; Ames, 1992). We also looked into heritage goals, which we believe are specific to learning subjects such as Hebrew language, which are connected directly to the heritage of the learner (Spolsky, 1989). We checked the academic efficacy of the learner, and his/her general satisfaction with the Hebrew language course. Past research found that boys showed a higher level of performance-approach goals and a lower level of mastery goals than girls (Nolen, 1988; Middleton and Midgley, 1997; Pajares and Giovanni, 2001). Another factor which should be taken into account is the possible effect of the gender composition of the classroom. Past research found that some girls are higher achievers in all-girl classes than in co-ed classrooms (Sadker and Sadker, 1994; US General Accounting Office, 1996). This study includes some participants who are learning in co-ed classrooms and others learning in segregated classrooms.

The most important innovation of this study was the cross-cultural aspect that demonstrates how religious values and classroom setting mediate the effect of gender on achievement and motivation when all are learning according to the same curriculum. In our research we chose to compare three subcultures within the Jewish North American community. All participants are Jewish students learning Hebrew language as part of their Jewish education. The three groups differ in their religious affiliation and class gender composition, which is strongly connected to their perceived gender roles. The first subculture, to which most of the participants belong, is comprised of the Reform and the Conservative communities. These communities, from their beginning at the end of the nineteenth century, saw equality between men and woman as a basic value for Judaism in the modern era (Meir, 1999). The second subculture is the Modern Orthodox community. This community is well integrated into the general American society, and as such, in their non-religious existence hold the same feministic views as the rest of the middle-upper white American class. However, in their religious activities there is a strong commitment to the traditional inequality between man and woman, which gives the men a much more dominant role in religious public life. Changes in the role of women in religious activities are happening, but this is done slowly and within the boundaries of Orthodox religious laws (Sheshar-Aton, 1999). One indication of this group's belief in gender equality is their support of co-ed education. The third subculture is that of the right-wing Orthodox community. In this community equality between men and woman is seen as 'anti-value'. The differences between the genders are seen as a basic Jewish value, which should be preserved (Fishman, 2000). One way in which it is preserved is through segregated education for boys and girls from aged 3 or 5 years till they complete their education. Our sample includes only the most liberal from this community. We hypothesized that the different gender roles in the three communities will be reflected in a different gender pattern in Hebrew language achievement and motivation. It should also be noted that as we move from the non-Orthodox groups to the right-wing Orthodox group there is a greater emphasis on religious/Jewish studies in the school setting in general and that in the majority of the schools Hebrew language is seen as part of the religious/Jewish curriculum of the school.

The purpose of our research was to examine the differences in attitudes towards Hebrew language learning and learning outcome (i.e., achievement) between girls and boys from three groups within the Jewish North American community. It was expected that we would find in general a higher level of performance and more positive attitudes towards learning Hebrew as we advance from the non-Orthodox group to the right-wing Orthodox group. We also expected to find a different gender gap pattern in performance and in attitudes in the three groups.

Method

Participants

The 735 students in this study were all North American youth, in grades 7–12, learning in Jewish schools. They were all using the same curriculum for learning Hebrew as a second language (NETA). There were 388 girls (52.79 per cent), 333 boys (45.31 per cent) and 14 who did not indicate gender (1.90 per cent). Twenty-seven of the students identified themselves as having no religious affiliation (3.6 per cent), 81 as Reform (11.2 per cent), 303 as Conservative (41.22 per cent) and 256 as Orthodox (34.83 per cent). Sixty-eight students did not indicate their affiliation (9.25 per cent). Out of the 256 students who identified themselves as Orthodox; 103 learned in a co-ed Jewish setting (40.23 per cent), and 153 learned in a segregated Jewish school (59.77 per cent), in which there are separate buildings for boys and for girls. However, the two buildings share the same curriculum and teachers.

Materials

Classroom goal structure + student's academic efficacy + evaluation of the learning situation

The survey as a whole concerns the student's experience from his/her learning situation. It consists of 24 items that focus on the classroom's goal structure (performance-avoidance, performance-approach, mastery, heritage). The survey utilizes a five-point Likert-type scale (from strongly disagree to strongly agree). The sub-scales referring to perception of classroom achievement goals were taken from the PALS survey of Carol Midgley et al. (2000). The α for these sub-scales ranged from 0.757 to 0.858. We added four items concerning the heritage goal ($\alpha = 0.78$) of the classroom and seven items in which the student was asked to rate his level of academic satisfaction from his/her Hebrew class ($\alpha = 0.82$). We also included in this questionnaire the Academic Efficacy sub-scale of Midgley et al. (2000). The sub-scale includes seven items. The alpha for this sub-scale is 0.89.

Learning outcome

A multi-choice, 60-item Hebrew language proficiency level test (Kobliner, not published) was given to the research sample. The tests are built of 60-items in increasing levels of Hebrew language proficiency, from the preparatory level, through the beginners, intermediate and up to the advanced level. The test examines only the level of Hebrew language proficiency in the area of reading comprehension. Both versions of the test were shown to be reliable (α of test #1 is 0.95, alpha of test #2 is 0.93).

Procedure

The students were tested during their regular class time in their classrooms towards the end of the school year. They were given the attitude survey at the middle of the school year.

Results

The objectives of the study were to:

1. Verify the connection between attitudes and achievement.
2. Assess the effect of gender on attitudes and achievement.
3. Examine differences in gender effect in different groups within the Jewish North American community.
4. Examine possible reasons for the different gender effects in the various groups.

For the analyses to follow, the Type I error rate was set at $p < 0.05$ unless otherwise stated. For the analyses of variance all *post hoc* tests were conducted using the Scheffe method.

Connection between attitudes and language achievement

In order to investigate which of the variables predict the student's level of Hebrew language proficiency we conducted a regression analysis. Three out of the four learning goals (Mastery, Performance Approach, Heritage), one attitude scale (Academic Efficacy) as well as religious affiliation and gender class composition, and gender were found to predict 21.4 per cent of variability on students' Hebrew language achievement scores. Results of the regression analysis are summarized in the following table. Thus, in the following analysis of attitudes towards Hebrew language learning we will take into account only the four attitude variables that were found here to be connected to language performance.

Effects of gender and group on language achievement and attitudes towards language learning

At the univariate level there was a significant effect of gender on achievement ($F(1,408) = 28.66$, $p < 0.0005$). Inspection means show that girls ($M = 31.30$) are significantly higher in Hebrew language performance than boys ($M = 24.82$).

Table 13.1 Results of regression analysis

	Adjusted R Square	**F Change**	**Df**	**Sig. F Change**
Academic efficacy	0.056	24.348	1390	0.000
Mastery	0.118	26.099	1389	0.000
Gender	0.158	21.503	1388	0.000
Religious affiliation + co-ed/seg	0.176	9.694	1387	0.002
Correctness of placement	0.192	8.447	1386	0.004
Performance Approach	0.205	7.161	1385	0.008
Heritage	0.214	5.476	1384	0.020

Table 13.2 Achievement test score by gender

	Number of students	**Mean**	**SD**
Girls	328	31.30	14.17
Boys	279	24.82	14.85

This tendency of girls to outperform the boys was found also when looking at placement of students according to level of learning material instead of score on a Hebrew language proficiency test. Also here we see that a higher percentage of boys are placed in the lower levels of language learning, whereas a higher percentage of girls are placed in the higher levels. The gap in placement between boys and girls was greater at the extremes. The correlation between gender and level of learning found here is higher than the correlations found between gender and language achievement because of a stronger tendency to place the girls in a learning level that is above their test score than to do the same with the boys.

Table 13.3 Learning level by gender

	Percentage of girls	**Percentage of boys**
Preparatory	10.8	22.9
Beginners	31.6	36.6
Intermediate	28.0	22.2
Advanced	29.5	18.3
Total	100.0	100.0

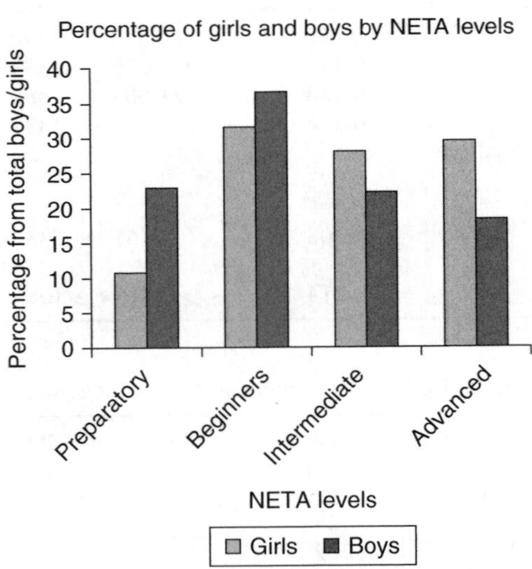

Figure 13.1 Learning material level by gender

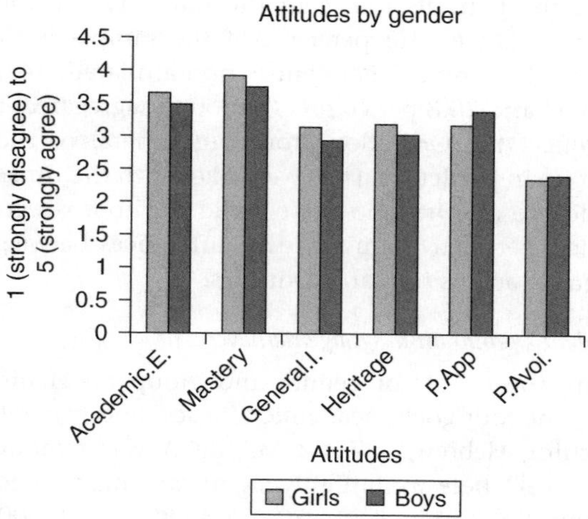

Figure 13.2 Attitudes towards learning Hebrew by gender

The differences of percentage of girls and boys by NETA level was found to be significant ($\alpha^2 = 25.45$, df = 3, $p < 0.0005$).

In addition, we also found corrections between gender and the way the student perceives his/her learning environment. Girls tend more than boys to believe in their ability to succeed in the Hebrew language course ($t = 2.355$, df = 445, $p < 0.019$), were more satisfied in general with the course ($t = 2.844$, df = 441, $p < 0.005$), and tended more to see the course as aimed towards mastery ($t = 2.748$, df = 420, $p < 0.006$) and connection to heritage ($t = 1.98$, df = 453, $p < 0.048$). Girls, compared to boys, tended less to see the course as aimed towards demonstration of ability ($t = -2.947$, df = 506, $p < 0.003$). Boys and girls did not differ in their tendency not to see the course as aimed towards avoidance of failure. These differences are similar to those found in learning in general (Nolen, 1988; Middleton and Midgley, 1997; Pajares and Giovanni, 2001) and in language learning in particular (Gardner and Lambert, 1972; Burstall, 1975; Politzer, 1983; Boyle, 1987; Oxford and Nyikos, 1988; Spolsky, 1989; Ellis, 1994; Huebner, 1995; Chavez, 2001).

Effects of student's group (religious affiliation + coed/seg) on gender differences in language achievement and attitudes towards learning

The gender gap in achievement and in attitudes described above might represent the majority culture in the North American Jewish community and overlook the patterns found in the minority subcultures within this community. Since 41.2 per cent of the sample is Conservative, 11.0 per cent is Reform, 3.7 per cent is non-affiliated, 14.0 per cent is Orthodox-co-ed and 20.8 per cent is Orthodox-segregated, these results might not reflect the Orthodox groups in general or the right-wing Orthodox group in particular. Thus we looked into the three subcultures (non-Orthodox-co-ed Orthodox-co-ed and Orthodox-segregated) and the connection between them and the differences between gender in Hebrew language acquisition and attitudes.

Effects of student's group on language achievement

To investigate the effects of gender and group on Hebrew language performance, the four goals, academic efficacy, and general satisfaction towards learning Hebrew, a 7×2 MANOVA was conducted. At the multivariate level, there were highly significant main effects for group (Wilks's Lambda = 0.855, $F(14, 804) = 4.688$, $p < 0.0005$) and for gender (Wilks's Lambda = 0.874, $F(7402) = 8.301$, $p < 0.0005$) and also for the interaction between group and gender (Wilks's Lambda = 0.873, $F(14,804) = 4.034$, $p < 0.0005$).

Inspection of the relationship between gender and group with scores and various attitude scales were assessed with analyses of covariance. To assess differences in performance in Hebrew language, performance was entered in analysis of covariance as the dependent variable with group and gender as covariates. The main effects for gender, $F(1600) = 6.30$, $p = 0.012$ and for group, $F(1601) = 19.43$, $p < 0.0005$ were found to be significant. Interaction for gender by group was highly significant, $F(1602) = 59.95$, $p < 0.0005$. Inspection of means indicate that among the Orthodox/segregated group the gap in performance between the girls ($M = 40.72$) and the boys ($M = 21.22$) was much larger then in the Orthodox/co-ed group (Girls: $M = 33.82$, Boys: $M = 30.40$) and in the non Orthodox group (Girls: $M = 27.03$, Boys: $M = 24.66$).

Table 13.4 Achievement test mean score by gender and group (religious affiliation/gender class composition)

	Non-Orthodox	**Orthodox Co-ed**	**Orthodox Seg.**
Girls	27.03 (sd 14.612)	33.82 (sd 11.476)	40.72 (sd 8.563)
Boys	24.66 (sd 15.619)	30.40 (sd 14.123)	21.22 (sd 12.156)

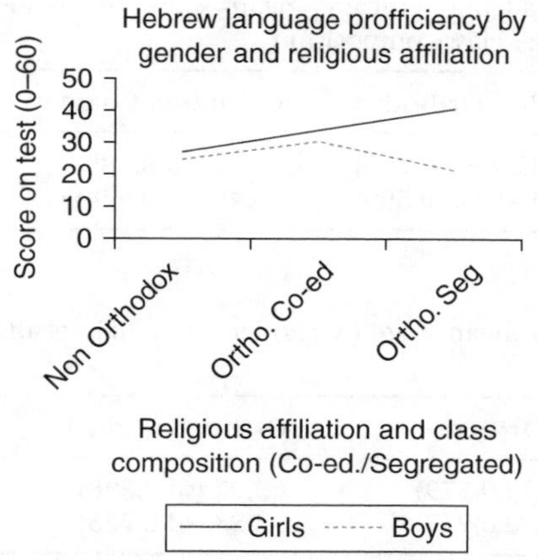

Figure 13.3 Achievement test mean score by gender and group (religious affiliation/gender class composition)

Effects of student's group on attitudes towards learning Hebrew

To assess differences in attitudes towards learning Hebrew language, the attitudes scores were entered into an analysis of covariance with the attitudes as the dependent variable and the group as the independent variable and gender as the co-variant. ANOVAs were also done on girls and boys separately. The following includes only the results for the attitude scales that were found to be related to achievement.

Academic efficacy. In an analysis of covariance with Academic Efficacy as the dependent variable and religious affiliation as the independent variable and gender as a covariate, the interaction of affiliation and gender was significant ($F(1465) = 6.44$, $p = 0.011$). ANOVAs were done separeately for the boys and the girls. A significant effect for religious affiliation was found only for girls ($F(1251) = 4.72$, $p = 0.031$).

Mastery. In an analysis of covariance with Mastery as the dependent variable and religious affiliation as the independent variable and gender as a covariate, the interaction of affiliation and gender was again significant ($F(1466) = 14.25$, $p < 0.0005$). ANOVAs done separately for the boys and the girls were not significant.

Heritage. In an analysis of covariance with Heritage as the dependent variable and religious affiliation as the independent variable and gender

Table 13.5 Academic efficacy mean score by gender and religious affiliation/gender class composition

	Non-Orthodox	Orthodox Co-ed	Orthodox Seg.
Girls	3.59 (sd 0.688)	3.74 (sd 0.785)	3.86 (sd 0.735)
Boys	3.45 (sd 0.896)	3.49 (sd 0.928)	3.57 (sd 0.773)

Table 13.6 Mastery mean score by gender and religious affiliation/gender class composition

	Non-Orthodox	Orthodox Co-ed	Orthodox Seg.
Girls	3.91 (sd 0.579)	3.97 (sd 0.598)	4.03 (sd 0.505)
Boys	3.78 (sd 0.717)	3.83 (sd 0.725)	3.40 (sd 1.035)

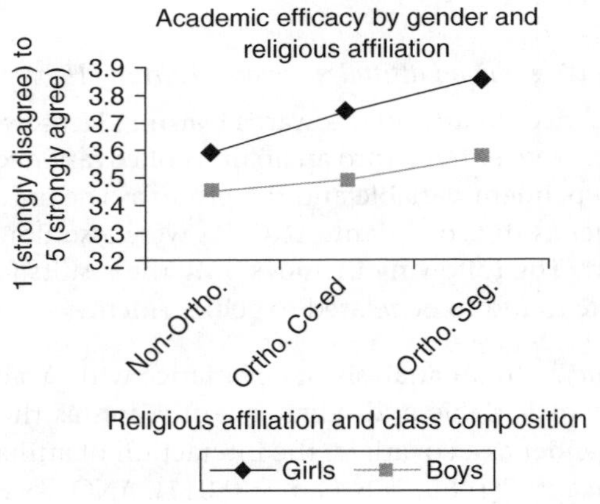

Figure 13.4 Academic efficacy mean score by gender and religious affiliation/gender class composition

as a covariate, the interaction of affiliation and gender was again significant (F(1464) = 9.99, p = 0.002). ANOVAs were done separately for the boys and the girls. A significant effect for religious affiliation was found only for girls (F(1251) = 6.44, p = 0.012).

Performance-approach. In an analysis of covariance with Performance Approach as the dependent variable and religious affiliation as the independent variable and gender as a covariate, gender was significant

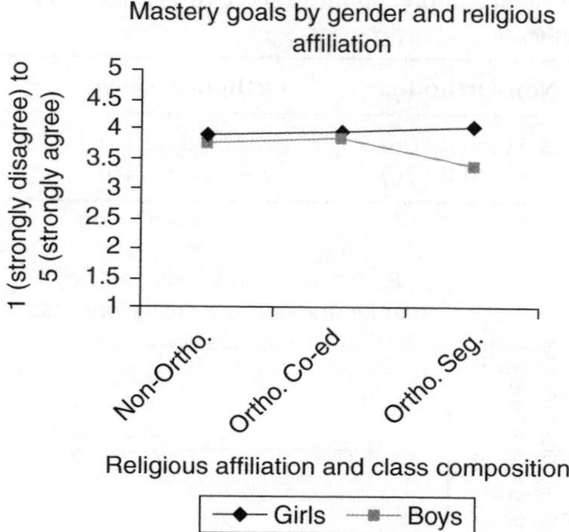

Figure 13.5 Mastery mean score by gender and religious affiliation/gender class composition

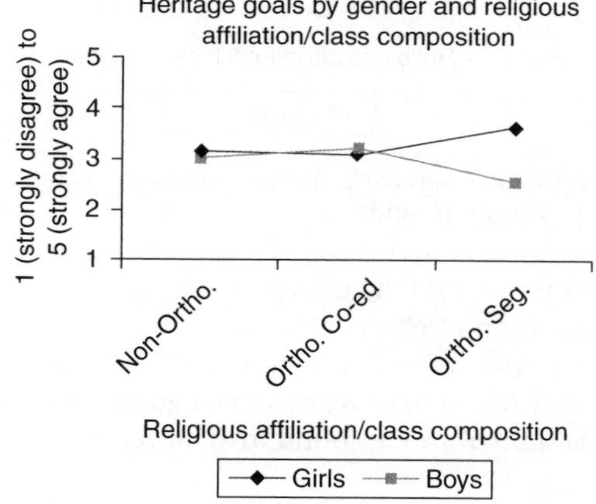

Figure 13.6 Heritage mean score by gender and religious affiliation/gender class composition

Table 13.7 Heritage mean score by gender and religious affiliation/ gender class composition

	Non-Orthodox	Orthodox Co-ed	Orthodox Seg.
Girls	3.14 (sd 0.850)	3.09 (sd 0.882)	3.68 (sd 0.772)
Boys	3.03 (sd 0.934)	3.21 (sd 0.946)	2.55 (sd 1.189)

Table 13.8 Performance-approach mean score by gender and religious affiliation/gender class composition

	Non-Orthodox	Orthodox Co-ed	Orthodox Seg.
Girls	3.21 (sd 0.806)	3.22 (sd 0.919)	3.11 (sd 0.730)
Boys	3.36 (sd 0.870)	3.45 (sd 0.716)	3.49 (sd 0.624)

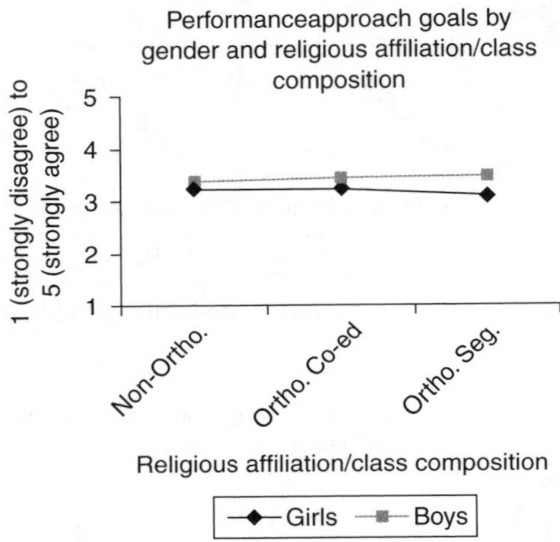

Figure 13.7 Performance-approach mean score by gender and religious affiliation/gender class composition

(F(1464) = 7.97, p = 0.005). ANOVAs done separeately for the boys and the girls were not found to be significant.

As seen above, two out of the four attitude scales (Mastery and Heritage) that were found to be related to language achievement showed a pattern that is similar to that found in achievement.

Discussion

This study found that the Orthodox/segregated (OS) group had a different gender gap pattern than the two other groups (non-Orthodox/co-ed = NOC, Orthodox/co-ed = OC). This different pattern was apparent in level of Hebrew language performance and in attitudes towards Hebrew language learning. In the OS group we see a significantly wider gap between boys and girls in favour of the girls in performance and in attitudes.

These differences between the groups can be explained by a combination of a cultural effect and learning environment effect (class gender composition), or perhaps the dominant effect of one or the other. As to the cultural effect, the OS group is the most right wing group of the three. This includes a basic belief that differences between genders are a Jewish value, which should be preserved. Males are the ones that hold public roles in religious public life and are the ones that are allowed and required to devote a large percentage of their Jewish education to the study of certain religious law texts (*Talmud*). Surprisingly, this dominance of boys in religious education and practice correlates in Hebrew language learning with a wider gap between boys and girls, in favour of the girls. This can be explained by the fact that the girl's education historically is under less religious restriction, which results in less resistance to innovative methods of teaching and teaching materials. In addition, since the girls are not allowed to learn *Talmud*, but they still are expected to devote half of each long learning day to Jewish studies, they are left with much more time and energy for studies such as Hebrew language. As to the learning environment effect, the success of the girls in the OS group can be attributed to the known positive effect of segregated settings on performance of girls (Sadker and Sadker, 1994; US General Accounting Office, 1996). It should be noted that past research on the effect of gender class composition did not find a negative effect of segregated settings on the achievement of boys. In order to determine the separate effects of the culture and the learning environment we would have needed to add to our sample a non-Orthodox segregated school. Such a school does not and probably would not exist since the idea of segregated education by gender contradicts the basic values of the American-Jewish Non-Orthodox communities. Another group, which could have enriched our understanding of cultural effects on gender differences, is the Ultra-Orthodox community. This group, which is more conservative in their religious values and practice and less open to American culture than our sample, has not agreed up to now to introduce the NETA curriculum in their schools, and thus are not included in this study.

It should be noted that although we also expected a larger gap between genders in the OC group, in comparison to the NOC groups, this was not found. This finding supports Fishman's (2000) claim that the Modern Orthodox North American community holds more liberal views then expected. It might also reflect the fact that once education is 'co-ed' boys and girls receive the same learning opportunities which can override cultural values of inequality.

Conclusion

The results of this research support the general known findings regarding the slight superiority of girls in second language learning as tested through written tests. It also supports the claims that girls have a more positive attitude towards the second language course, and are aimed more towards mastery of the language and less towards demonstration of ability. This study also succeeded in applying the achievement goal theory to second language learning and to show that there is a connection between attitudes and achievement.

The innovation of this research is that the general pattern above, which was found in research projects in the past, seems to be a correct reflection of the majority culture in North America, but overlooks gender patterns found in minority subcultures. This different gender pattern of achievement and attitudes towards second language learning was found here in the most right-wing Orthodox group in the sample. This group holds values regarding gender roles in religious practice and education which differ from the values that are dominant in the majority culture. Surprisingly, in those communities, which generally give boys more dominant roles in the process of religious socialization, we find a wider gap between boys and girls in language performance and attitudes towards learning **in favor of the girls**. This might result from different characteristics of this group, such as more time and energy that girls devote to language learning or from the overriding effect of gender segregated education on the achievement and attitudes of boys and girls. This study is the first cross-subcultural study with school-aged students which supports poststructuralist approaches, viewing gender as a social, historical and cultural construct, which mediates between culture and language behavior (Gal, 1991; Piller and Pavlenko, 2001). It also succeeds in showing that gender mediates between culture and language via attitudes.

References

Ames, C. (1992). Classrooms: Goals, structures, and student motivation. *Journal of Educational Psychology* 84, 261–71.

Bacon, S. M. (1992). The relationship between gender, comprehension, processing strategies, and cognitive and affective response in foreign language listening. *Modern Language Journal* 76, 160–78.

Baker, S. C., and MacIntyre, P. D. (2003). The role of gender and immersion in communication and second language orientation. In: Z. Dornyei (Ed.), *Attitudes, orientations, and motivations in language learning*. Nottingham: University of Nottingham.

Bonvillain, N. (1997). *Women and men: Cultural constructs of gender*. 2nd edn, Englewood Cliffs, NJ: Prentice Hall.

Boyle, J. (1987). Sex differences in listening vocabulary. *Language Learning* 37, 273–84.

Brecht, R. D., and Robinson, J. L. (1993). Qualitative analysis of second language acquisition in study abroad: The ACTR/NFLC project. *NFLC Occasional Papers*.

Burstall, C. (1975). Factors affecting foreign-language learning: A consideration of some relevant research findings. *Language Teaching and Linguistics Abstracts* 8, 5–25.

Butler, R. (1987). Task-involving and ego-involving properties of evaluation: Effects of different feedback conditions on motivational perceptions, interest and performance. *Journal of Educational Psychology* 79, 474–82.

Chastain, K. (1970). A methodological study comparing the audio-lingual habit: Theory and the cognitive code learning theory – A continuation. *Modern Language Journal* 54.

Chavez, M. (2001). *Gender in the language classroom*. Boston, MA: Heinle & Heinle.

Dweck, C. S., and Elliott, E. S. (1983). Achievement motivation. In: P. Mussen and E. M. Hetherington (Eds). *Handbook of child psychology*. New York: Wiley.

Dweck, C. S., and Leggett, E. L. (1988). A social-cognitive approach to motivation and personality. *Psychological Review* 95, 256–73.

Eckert, P., and McConnell-Ginet, S. (1992). Think practically and look locally: Language and gender as community-based practice. *Annual Review of Anthropology* 21, 461–90.

Ellis, R. (1994). *The study of second language acquisition*. Oxford: Oxford University Press.

Fishman, S. B. (2000). *Jewish life and American culture*. Albany, NY: State University of New York Press.

Gal, S. (1991). Peasant men can't get wives: Language and sex roles in a bilingual community. *Language in Society* 7 (1), 1–17.

Gardner, R. C., and Lambert, W. E. (1972). *Attitudes and motivation in second-language learning*. Rowly, MA: Newbury House.

Holmes, J., and Meyerhoff, M. (1999). Communities of practice in language and gender research. Special Issue of *Language in Society* 28 (2).

Huebner, T. (1995). A framework for investigating the effectiveness of study abroad programs. In: C. Kramsch (Ed.), *Redefining the boundaries of language study*. AAUSC Annual Volumes. Boston, MA: Heinle & Heinle (pp. 185–217).

Klee, C. (1987). Differential language usage patterns by males and females in a rural community in the Rio Grand Valley. In: T. Morgan, J. Lee, and B. VanPatten (Eds), *Language and language use*. Lanham, MD: University Press of America.

Lynch, A. (2004). The relationship between second and heritage language acquisition: Notes on Research and theory building. *Heritage Language Journal* 1.

Maehr, M. L. (1989). Thoughts about motivation. In: C. Ames and R. Ames (Eds), *Research on motivation in education: Goals and cognitions*. Vol. 3, New York: Academic Press (pp. 299–315).

Middleton, M. J., and Midgley, C. (1997). Avoiding the demonstration of lack of ability: An underexplored aspect of goal theory. *Journal of Educational Psychology* 89, 710–18.

Midgley, C., Maehr, M. I., Hruda, L. Z. et al. (2000). *Manual for the patterns of adaptive learning scales*. Michigan: www.unich.edu

Nicholls, J. G. (1984). Achievement motivation: Conceptions of ability, subjective experience, task choice, and performance. *Psychological Review* 91, 328–46.

Nolen, S. B. (1988). Reasons for studying: Motivation orientations and study strategies. *Cognitive and Instruction* 5, 269–87.

Nyikos, M. (1990). Sex related differences in adult language learning: Socialization and memory factors. *Modern Language Journal* 3, 273–87.

Oxford, R., and Nyikos, M. (1988). Variables affecting choice of language learning strategies by university students. *Modern Language Journal* 73, 291–300.

Pajares, F., and Giovanni, V. (2001). Gender differences in writing motivation and achievement of middle school students: A function of gender orientation. *Contemporary Educational Psychology* 26, 366–81.

Piller, I., and Pavlenko, A. (2001). Introduction: Multilingualism, second language learning and gender, In: A. Parlenko, A. Blackledge, I. Piller, and M. Teutsch-Dwyer (Eds), *Multilingualism, second language learning and gender*. Berlin: Walter de Gruyter.

Politzer, R. (1983). An exploratory study of self-reported language learning behaviors and their relation to achievement. *Studies in Second Language Acquisition* 6, 54–67.

Sadker, M., and Sadker, D. (1994). *Failing at fairness: How America's schools cheat girls*. New York: Touchstone.

Shiue, C. (2003). English foreign language acquisition among college students-listening skills. *Hwa Kang Journal of TEFL*.

Sole', Y. (1978). Sociocultural and sociopsychological factors in differential language retentiveness by sex. *International Journal of the Sociology of Language* 17, 29–44.

Spence, J. T., and Helmreich, R. L. (1983). Achievement-related motives and behaviors. In: J. T. Spence (Ed.), *Achievement and achievement motives: Psychological and sociological approaches*. San-Francisco, CA: Freeman (pp. 7–74).

Spolsky, B. (1989). *Conditions for second language learning: Introduction to a general theory*. Oxford: Oxford University Press.

US General Accounting Office. (1996). *Public education: Issues involving single-gender schools and programs*. Report to the Chairman, Committee on the Budget, House of Representatives: Washington, DC.

Zentella, A. C. (1997). *Growing up bilingual*. Malden, MA: Blackwell Publishers.

מאיר, מ. (1999) . "אלוהי אברהם ושרה" מעמד האשה ביהדות הלא אורטודוקסית. מתוך ד.י. אריאל, מ. ליבוביץ וי. מזור(עורכים), ברוך שעשני אשה? האשה ביהדות- מהתנ"ך ועד ימינו. תל אביב, ישראל: ידיעות אחרונות וספרי חמד.

ששר אטון, נ. (1999) . " גילוי חדש של הרצון האלוהי" השפעת הפמיניזם על הציונות הדתית". מתוך ד.י. אריאל, מ. ליבוביץ וי.מזור (עורכים), ברוך שעשני אשה? האשה ביהדות- ועד ימינו. מהתנ"ך תל אביב, ישראל: ידיעות אחרונות וספרי חמד.

14
Speaking Our Gendered Selves: Hinduism and the Indian Woman*

Kalyani Shabadi

This chapter explores the construction of the Hindu Indian woman and how language formalizes and ritualizes a particular way to be. It approaches the questions of how a gendered self is linguistically shaped by religious traditions and shows how the ascription of positive and negative values in religion leads to gender discrimination and asymmetry.

The topic of the self dealing with personhood, gendered identity, the body and selfhood of women has become a major issue in the feminist philosophy. Feminist theorists have interpreted gender as the experience of sexed embodiment, a set of internalized norms, a set of traditional roles, a performance, a social position or class. A constructive view of gender like that of Butler (1990) and others proposes that ways of talking and behaving that are associated with gender are a matter not of identity but of display. Butler says: 'There is no gender identity behind the expressions of gender; ... identity is performatively constituted by the very 'expressions' that are said to be its results.' For constructionists, 'gender is doing, not being', and thus they emphasize that each individual must constantly negotiate the norms, behaviours and discourses that define masculinity and femininity for a particular community. In this context, we will discuss how language is used in a speech community in performance that indicates the social identities of men and women. How do particular linguistic practices contribute to the production of people as 'women and men'?

* I give my sincere thanks to Dr Rajashekhar Shabadi for the data in Kannada and Dr Thomas Chacko for the data in Malayalam.

This chapter explores the construction of gender identities in Indian society with relation to Hinduism. It shows how gender identity is formalized and ritualized through language and how different socio-religious contexts are intertwined with understandings of gender dynamics. It describes how a gendered self is shaped by religious traditions, which is reflected in the language of the society. It shows how gender identities have been invented and valued in different socio-religious and regional contexts in India. I focus on religious traditions attributed to men and women in India and discuss the dis/advantages of religious practices that have been rooted in the Hindu culture in which we conceive of belief systems in contemporary society. For this purpose, I consider a natural-gender language like Oriya along with some examples from some other Indian languages like Hindi, Kannada and Malayalam.[1]

This chapter gives a brief description of the Indian socio-cultural setup, and discusses how selfhood is realized in Hinduism. I then present how, as a product of social reality, language reflects the socio-cultural behaviour of a community. I discuss linguistic, psychological and sociological issues in the use of gendered terms in language and show how generic masculine terms, taboo expressions and certain other words used in daily life indicate the gender bias in the society. Finally, I conclude with some plausible solutions to bring social change.

Men and women in the socio-cultural set-up

Men and women are socially different because of the different social roles (along with the biological roles, of course) imposed upon them by the society. If we consider the status of men and women in Indian society, we find that the status of women is far below the status of men. In Indian society, the gender-based division of labour is rigid among different caste and class groups. It is observed that middle-class as well as higher-caste people have more gender segregation.

According to the official law in India, women have the same rights as men. There are places reserved for women in the professional and educational fields. A number of government organizations as well as non-government organizations like *mahiLaa samaaja* 'women society', *mahiLaa kendra* 'women centre' are designed to work for the welfare of women in the society. There is a strong group protest against alcoholism, as men usually squander money in drinking irrespective of their financial condition, and they protest against discriminating family laws, rape and dowry laws and violence too. A number of magazines give enough space for articles on women's issues. In spite of all these legal rights of women, there is much discrimination against women. Dowry-murders and violence against women are common in many parts of India.[2] Women are confined to household and reproductive work. Their mobility is restricted both in rural and urban areas of India. Their position is further disadvantaged due to lack of property rights. Patrilineal inheritance and patrilocal residence makes women dependent on men.

More often culture has been seen closely associated with religion, and other social systems such as marriage, eating habits, way of living; and this association has brought tremendous impact on the lives of women. The Hindi word *purdah/ghungaT* 'veil', or the seclusion and segregation of women, is a cultural norm on the Indian subcontinent, although it has been reduced these days. Beginning at puberty, *ghungaT* controls the interaction between females and males. Cultural constraints on women writers make it more or less taboo for women to tackle certain topics like sex, religion and politics. Cultural constraints on women are often rooted in religion. However, Hinduism has become part of the culture within India and if one deviates from these accepted practices one is looked down upon by the society. As a product of social reality, language reflects the socio-cultural behaviour of its speakers, it reflects the thoughts, opinions, attitudes and culture of its users. Gender bias is reflected in various parts of life, including language, in Indian society.

Hinduism and 'selfhood'

In India, Hindu society has often been divided on the basis of caste, region, gender, language and belief. Hinduism recognizes social and economic inequalities among human beings as inevitable constituents of society because of the individual differences in the nature of their *karma* 'one's own doings', and often it is related to the *karma* of one's previous birth. If a person is rich, happy, good looking or born into a higher caste, it is because of his *karma*; and if someone is suffering, poor or born in a low caste, he has to blame himself and his previous actions. Each individual who lives in this world is a continuation of his/her past and is fully responsible for his or her present reality. This is the 'selfhood' assigned to each individual by God. Therefore, none can blame others for his/her own suffering. Brought up in a society that is organized on the basis of a caste system and on social values that are centred around the concept of *karma*, an average Hindu is more obsessed with the problems of his evolutionary impasse that is his selfhood.

Religion works at many different levels in the lives of women. At one level, religion has often been used for reinforcing patriarchy and the existing caste and class divisions in society. *Pativrata* 'devoted to one's husband' is the ideology or specific *dharma* 'religion' of the Hindu wife by which women accept and aspire to chastity and wifely fidelity as the highest expression of their selfhood.

Right from the time when she is a little girl, a female is brought up with the idea that marriage and motherhood are her final legitimate destinations, even at the expense of sacrificing and neglecting her own needs and aspirations. The dread offence a woman is guilty of is not to bear children. Her life is reduced to violence, rejection and misery. The childless woman, and not the man, is harassed, beaten, insulted, threatened by remarriage of her husband or deserted. She is called names like *banjh* 'barren' or *sookhi kokh* 'dry womb' in Hindi. All these beliefs seem to be designed for a man's advantage.

In traditional societies, specific traits were assigned to men and women. Indian culture had set up strict social norms for the sexes. Attributes like 'knowledge', 'mental ability', 'determination', 'firm decision', 'faith' and 'confidence' were ascribed to males, whereas 'shyness', 'ignorance', 'fear', 'timidity' are the qualities ascribed to females. Man was entitled to get knowledge, wisdom and strength whereas woman lacked all that, thus indicating the superior and inferior status of men and women respectively. Education, visiting abroad, possession and demonstration of valour, earning fame, participation in decision-making meetings and giving charity were the sole privileges and goals of a man and no such privileges were available to women. Women and *Sudras* 'low-caste people' were declared to be unfit for study of the *Vedas*.

Tradition does not allow a woman to remarry or break the marriage, but allows a husband to abandon his wife. A divorced woman is labelled as *chaaDuri* in Oriya, a very derogatory term that is seldom used for a man. The matter of description of human qualities is based, for instance, on this double standard. A bold man is interpreted as *rokaDaa* 'courageous', but a bold woman is *rokaDaa* 'aggressive' in Oriya. A woman can be discarded by her husband easily, as in Kannada he is the *yajamaana* 'the one who owns her.'

In law, in customary practice and in cultural stereotypes, women's selfhood has been subordinated and in certain cases such subordination is accepted by law. A woman's personhood is absorbed into that of her husband when she marries. The wife assuming her husband's name/surname symbolizes this revocation of her separate identity. Irigaray (1993) has rightly said that man seems to have wanted, directly or indirectly, to give the universe his own gender as he has wanted to give his own name to his children, his wife, his possessions. Women are defined by their relation to men. A married woman is usually referred to or addressed as 'Mrs + husband's name' or the 'mother of + the child's name'. Women adopt the family name of their husbands, children are given the family name of their male parent. Male offspring are considered as heirs. A query about a child such as *kaahaa pua/jhia* 'whose son/daughter' in Oriya is referred to by stating the name of the child's father (not the mother). She loses her right to property as her husband is entitled to control her earnings. The highest *dharma* 'religion' of woman is to worship her husband. In childhood she is in the custody of her father, in youth she is in custody of her husband, and in old-age she is in the custody of her sons. In Oriya, the popular saying goes *binashraye nabanchanti kabitaa, banitaa, lataa*, that is, 'poetry, creeper and women need support to survive'.

The son is seen as the *uttaraadhikaari* 'heir' to the family name and fortune whereas the daughter is seen as a liability, whose parents must find her a husband to secure her future. These beliefs pervaded Indian society leading to the dowry system, where the family of the woman paid her husband to 'take her off their hands'. Slowly the Hindu religion incorporated these principles through commending the birth of a male. Women are often blamed or shamed for giving birth to a daughter. 'Son preference' reveals the dynamic by which the status of women can have a profound, even fatal, impact on the health of women and girls. There are also many reports of female infanticide which occurs because the raising of a girl can often be a financial burden. The problem of female infanticide and its resulting gender imbalance is deeply rooted in India's subjugation and subordination of women.

In Hinduism the woman's role is often seen within the context of the family. Often women are killed for having sexual relations outside marriage, choosing their own husbands against parental wishes or seeking a divorce. A husband (as well as his kin) could at any time accuse his wife of infidelity and she would have to pay with her life for her husband's or elders' mere suspicions. Even goddess Sita had to pass through the fire ordeal after her return from Sri Lanka, as she was required to by her spouse *Ram*, the 'ideal husband' of the 'Hindus'.

Devadasi 'god's slave$_{\text{Fem}}$' – the practice of dedicating young low-caste girls (mahars, Mangs, Dowris and Chambhar) at childhood to a god and their initiation into prostitution when they attain puberty continues to thrive in Karnataka, Andhra Pradesh and other parts of South India. The girls from poor families are married to God Krishna and are sold after puberty at private auctions to a high-caste master who initially pays a sum of money to the families of the girls. This link between religious culture and child prostitution is largely due to social backwardness, poverty and illiteracy.

Religion has also prescribed many dietary restrictions on women, especially on widows. Widows are not supposed to eat good food items like non-vegetarian food items, onion, garlic and so on. The idea is that without their husbands around, they are not supposed to eat *tamasika* 'aphrodisiac' food as it may stimulate their sexual desires. Widows are not supposed to wear coloured clothes other than white, or wear make-up to look gorgeous, they are not supposed to attend auspicious ceremonies like marriages. The idea is that without the man in her life, the woman should lead a life of renunciation and spirituality. Similar restrictions are there for a divorcee also. A divorced woman cannot take active part in the rituals of a marriage ceremony, and for that matter, nor in any auspicious occasions. On the other hand men are exempted from any such restrictions.

The ideology of the good wife and the good mother demands that the woman eats last and eats what remains after her children and husband are fed. In situations when there is limited availability of food, sons are fed before daughters. Even if women do the earning, their incomes are placed in the hands of male decision-makers in the family. The woman learns early her lessons of sacrifice by giving priority to the requirements and likings of other members in the family. As Khokle (1995) puts it, the (ideal) husband is devoted to his work and that, for him, his duty is as dear as his life, whereas for the wife, who is confined to her home, her sole life is her husband.

However, in certain cases women are given importance and respect as well. Unlike many other religions, in Hindu religion female deities are very important. In India, it is said that there is a goddess in every village *graama devati* 'village deity', known as the protector of the villagers from any sort of evil. Hindu philosophy interprets the goddess as the *Shakti*, or cosmic energy, and therefore the most immediate creative or destructive force, to be thanked or placated. Many of the manifestations of the goddess *Kali* or *Durga* are capricious or violent, and she is often seen as a warrior who destroys demons on her own. As *Mariammaa* in Kannada or *ThaakuraaNi* in Oriya, she used to bring smallpox, and she is still held responsible for diseases of the hot season.

The women's role is not simply to cook food and keep the home running but also to make sure the family keeps the religious rituals and celebrations important to the Hindu faith. As such, women have a very important role to play and certain rituals like *yagna*, wedding ceremonies, funeral rites cannot be performed without women. So it could be said that women have been given a status and respect in spiritual matters. The concept of universe as union of the female (*prakruti*) and male (*purusa*) principle gives the female equal status with the male. Women like Indira Gandhi, Sonia Gandhi, Jayalalitha and so on have gone to the top position in their careers irrespective of the religious category or caste they belong to.

Gender and language

In spite of the fact that Oriya is a natural gender language (Sahoo, 2003), (Beames, 1966),[3] (Priestly, 1983),[4] gender bias is reflected in the language of everyday life in Oriya society. A comparison of the qualities ascribed to both the sexes in and across languages would reveal actual values put on males and females in society.

Generic masculines

Generic masculine words like *aadimaanaba* 'prehistoric man', *purba purusa* 'forefather', *saata purusa* literally 'seven man' meaning 'seven generations' assume the inherent superiority of the male over the female. As Clark and Clark (1977) puts it, the unmarked category represents both maleness and femaleness, while the marked represents femaleness only.

Certain masculine adjectives are used generically, for example *budhiaa* 'intelligent', *kuhaaLiaa* 'outspoken', *gyaani* 'wise', *chaalaaka* 'clever', *chatura* 'intelligent', *raagi* 'angry', *bekaara* 'unemployed', *lobhi* 'greedy', *bokaa* 'fool', *krupaNa* 'miser', *chandaa* 'bald', *priya* 'dear', *siaaNiaa* 'clever', *nicha* 'mean', *amaaniaa* 'disobedient', *dhani* 'rich', *gariba/daridra* 'poor' and so on. Generic masculine adjectives like *dhani* 'rich', *gariba* 'poor', *bad-kharchi* 'spendthrift', *daani* 'donor', *krupaNa* 'miser' and so on show the male-oriented financial dealings of the society. It is usually the man who earns the livelihood, the woman takes care of it only. Man is responsible for the financial condition of the family. In most cases, money is earned and spent by the male members of the family. Property is inherited in the name of the man. The financial status of a woman is usually based on the financial status of her husband.

Certain male nouns like *lakhyapati* 'owner of lakhs of rupees', *koTipati* 'owner of crores of rupees', *niyutapati* 'millionaire', *jamidaara* 'landlord', *saahukaara* 'money-lender', *kiNaaLi* 'buyer', *bikaaLi* 'seller' and so on are also used generically.

Some derived nominals (deverbal nouns) associated with certain professions and certain types of activities, which carry masculine gender, are also always used generically as women are not supposed to do these kinds of jobs. For example *bulaa bikaaLi* 'hawker', *paahaaDa chaDhaaLi* 'hill climber', *ghaasa kaTaaLi* 'grass mower', *kaaTha kaTaaLi* 'wood cutter'. Terms like *bikaaLi* 'seller' and *kiNaaLi* 'buyer' are always masculine, because it is always men who are involved in buying and selling activities.

Quantifying expressions like *kie/kehi jaNe* 'someone', *jaNe loka* 'a man' are used generically. However, occasionally one finds the use of feminine gender in a generic sense. For example, one has the *maatrubhaasaa* 'mother tongue', but has no *pitrubhaasaa* 'father tongue'. Here, *maatrubhaasaa* 'mother tongue', is used in a generic sense (the language which a child acquires first in childhood).

Taboo expressions and gender terms of abuse

Taboo expressions reflect the gender bias of Oriya society. Certain words of abuse do not have a female counterpart and they are usually used generically. For example, *udhata* 'proud', *bajaari* 'loafer', *dhurta* 'cunning', *abibeki* 'having no conscience', *agyaani* 'unwise', *amaNisa* 'not a human being', *murkha* 'dull', *gajamurkha* 'dull to the core', *ghusuri* 'pig', *amaaniaa* 'disobedient'. Man only can be qualified or disqualified for all these things.

Taboo expressions concerning 'death' also reflects the gender bias of the society. Terms like *baaDikhiaa* 'die of cholera', *saapakhiaa* 'be bitten by a snake', *baghakhiaa* 'be killed by a tiger', *bhenDaabansiaa* 'all the people/members of the family should die young', *gajaa bayasiaa* 'die in young age', *adhaa bayasiaa* 'die early' and so on are used exclusively for men, which shows the importance attached to a man's life. It gives the impression that such terms are not used for women as there is little importance attached to the lives of women in the society. Similarly, expressions like *pua khaai* 'to be sonless' (literally, 'the woman who eats her own son') or *ghaitaa khaai* 'to be a widow' (literally, 'the woman who eats her own husband') are used as terms of abuse for a married woman. But terms of abuse for the expression 'to be a widower' are rarely employed for men. Also, *jhia khaai/khiaa* 'to be daughterless' hardly occurs as a term of abuse. And again this indicates that very little importance is attached to the life of a female child. The male term *amaNisa* 'worthless/hopeless person' is used in a generic sense, which includes both male and female.

Terms like *raanDa/bidhabaa* 'widow', *daari* 'whore', *besyaa* 'prostitute', *baanjhuNi* 'barren', *chhaaDuri* 'divorcee', *bedhei* 'having pre-marital affairs' are usually used as words of abuse, and are applicable only to women. There is hardly any masculine counterpart of these words. The reason would seem to be that in a male-dominated society like that of

Oriya, it is believed that misfortunes which arise in the course of marriage, like barrenness, widowhood, divorce, along with socially unacceptable conduct such as prostitution, pre-marital affairs are events or acts associated only with the female gender. A word like *rakshitaa* 'kept woman'/'mistress' does not have a masculine counterpart. Words like *DaahaaNi* 'witch', *kuLaTaa* 'having extra-marital affair', *patitaa* 'fallen woman' (usually used for women having extra-marital relationships) do not have masculine counterparts as these terms are not considered to be serious misconducts if committed by men.[5] So, there is no name or word for a man for doing the similar thing.

These taboo terms have some emotional effect on the hearer, since these terms diminish his/her status as a human being (Agyekum, 1996).

Asymmetry in lexical meaning with reference to gender

Asymmetry in lexical meaning is linked to perception of gender roles in a socio-cultural setting. In Oriya, certain words have been lexicalized along gender lines.

In certain professional as well as other fields of life, only men are supposed to work and thus one does not find female counterparts. For example *kaarigara* 'carpenter', *bindhaaNi* 'smith', *sainika* 'soldier', *daarogaa* 'constable', *raasTrapati* 'president', *jaguaaLa* 'watchman'/'security guard', *Daaka piana* 'postman', *laainmyaan* 'electrician', *niaan libhaaLi* 'fireman', *byabasaayi* 'businessman', *saapuaa* 'snake-charmer', *myaanejara* 'manager', *raajyapaaLa* 'governor', *rakhyaka* 'saviour', *draaibhar* 'driver', *Dakaaeta* 'dacoit', *dasyu* 'robber', *jaLadasyu* 'pirate', *mukhiaa* 'leader', *haLiaa* 'labourer', *graahaka* 'customer', *mahaajana* 'businessman' or 'money-lender', *jamidaara* 'landlord', *chelaa* 'follower', *majuriaa* 'day-labourer', *roseiaa/pujaari* 'cook' and so on. Although it is the woman who always cooks at home throughout her life, there is no term for a female cook! One also does not find a female cook in a hotel or in a public place in Orissa and for that matter in India too.

Conventionally, women are not supposed to smoke or drink alcohol. So terms like *maduaa* 'drunkard', *biDiaa* 'smoker', *ganjeiaa* 'hashish-addict', *aapuaa* 'opium-eater' and so on are always in the masculine gender and are used exclusively with reference to men. Even if a woman were to qualify as a drunkard, smoker or drug-addict, a periphrastic expression would be used, but never the masculine term in the language.

Terms related to knowledge – *panDita* 'learned person', *murkha* 'stupid', *gyaani* 'wise', *agyaani* 'dull' and so on are assigned to men only. Nouns like *harijana* 'a person from low caste', *girijana* 'a tribal man', *haLiaa* 'labourer', *mahaaprabhu* or *mahaapru* 'god'[6] and so on are male terms which do not have a feminine counterpart. Man has not forgotten to assign his own gender to his belief system, of 'god' even. So, although gods and goddesses are available in the Hindu belief system of god, still *ThaakuraaNi* 'goddess' or *bhagabati* 'goddess' is never used generically. Terms with reference to god, such as, *ishwara* 'god', *parama pitaa* 'the great father', *srusti kartaa* 'the creator' are treated as masculine. *Baapa ghara* 'father's house' is used generically for 'parents' house', and there is no term like 'mother's house'.

For certain compound nouns, Oriya strictly follows a $[N_{Masc} + N_{Fem}]$ order and the reverse order is not possible. For example, *bhaai bhauNi* 'brother sister', *baapaa maa* 'father mother', *pua jhia* 'son daughter', *swaami stri* 'husband wife', *bara kanyaa* 'bridegroom bride', *bhaai bhaauja* 'brother sister-in-law', *ajaa aai* 'maternal grandfather and grandmother', *maamu maain* 'maternal uncle and aunt', *daadaa khuDi* 'paternal uncle and aunt' and so on.

Eka patni vrata 'to have a single wife' is associated with the religious epic *Ramayana* where lord Rama had taken an oath to have a single wife in his life. This indicates that the practice of polygamy was allowed to men but not to women, so a word like *eka pati vrata* 'to have a single husband' is never found in the literature.

Words borrowed from English like *polis* 'police', is always considered to be masculine, and *nars* 'nurse' is assumed to have feminine gender on the default basis. The idea behind the perception of the respective gender roles of these terms is that the police are portrayed as the maintainers of discipline in the society and so it has to be a man. On the other hand, a nurse, who takes care of patients/people, has to be a woman. This relates to the fact that man is strong and powerful, while woman is loving and mother-like. So, one also finds words like *maaikinaa polis* 'female police', and *anDiraa nars* 'male nurse' as a feminine or masculine counterpart of the terms, respectively.

It is obvious that the male counterpart of *sati* 'pious woman' does not exist in the language as the piety of a man is an ignorable factor, and in *sati* 'rites' a man has never followed his dead wife in to the fire. Similarly, certain expressions do not have any female counterpart, for example *manara maNisa* (literally, 'mind's man' or 'the man of one's heart'), *pati parameswar* 'husband is the sole God'. As Hasan (1997) explains, the husband is the sole god for his wife, *iha kaaLa para kaaLara devataa* 'the god of her life in this life as well as in her afterlife'.

Aggressive and obscene idioms like *puliyaadi mon* 'son of a whore' in Malayalam and *raanDipua anantaa* 'son of a widow' in Oriya shows the importance of a father for a child. Expressions of benediction like Sanskrit *putravati bhava* 'may you have sons', but not *kanyaavati bhava* 'may you have daughters' show the preference for sons over daughters.

The inferior status of a woman is further revealed in the non-reciprocal usage of the forms of address. In Oriya, the three variants of 'you': *tu, tume* and *aapaNa*, which correlate with the three levels of honour and intimacy reflect the gender-bias of the society (Sahoo, 2003). For example, a child uses the *tume* 'you$_{[+\text{honorific}]}$' variant to address his/her father, but uses the *tu* 'you$_{[-\text{honorific}]}$' variant for the mother. Like Oriya, Hindi which has a three-tier system of second-person pronouns (*tu, tum* and *aap*) and Kannada, which has a two-tier system of pronouns (*ni:nu* and *ni:vu*) also follows a similar pattern. In Kannada, a husband generally addresses his wife by name or he uses a non-honorific pronoun, namely, *ni:nu* 'you$_{[-\text{honorific}]}$', while a wife uses *ni:vu* 'you$_{[+\text{honorific}]}$' or *ri* 'a term of respect'. While a husband uses a non-honorific or less honorific reference pronoun *avaLu* 'she$_{[-\text{honorific}]}$' to refer to his wife, she refers to him by a honorific pronoun *avaru* 'he$_{[+\text{honorific}]}$'. As non-naming denotes respect in many cases, she follows this pattern of address as well as reference. Assuming a superior status in the society, a man commands his wife by using non-honorific singular imperatives like *ba:* 'come', and *koDu* 'give'. However, a woman uses the honorific form like *banri* 'come$_{[+\text{honorific}]}$', *koDi/koDri* 'give$_{[+\text{honorific}]}$'.

So, linguistic practices reflect all these gender related hierarchies prevalent in the society and thus demonstrate an unmistakable asymmetry between men and women.

Conclusion

The nature of gender differences in Oriya clearly reflects the social and cultural factors responsible for the discrimination prevalent in Orissa. Similar vocabulary differences on the basis of gender bias exist in other Indian languages. These differences are the result of the differences in the position and the status of the two genders in society. When these conditions change, the differences are bound to modify. Hence the results of this research can be easily attested in many of the Indian languages.

Plausible solutions would be gender empowerment or gender development. Such an attempt has a role to play in strengthening the position of women in society. There is a need to remove the gender bias in language (i.e., to bring oral equality among men and women), which gets translated into other spheres of social, economic and political activities, to create awareness among both men and women about the consequences of gender bias.

Many non-governmental organizations are trying to bring broad social change. These programmes include assistance to mothers with childcare support, projects to improve women's access to education, and to educate women about their subordination within a patriarchal society. Along with these steps, it is necessary to implement policies and programmes for the promotion of equality for women in political, legal, economic, educational and social spheres. Efforts should be made towards shifting the social attitudes and long-standing traditions that contribute to gender-structured social beliefs.

However, we shouldn't assume that being a woman means never disagreeing, never contesting or never asking for evidence. Similarly, being a man does not always mean being aggressive and disagreeable. A typical woman is sometimes empathic and compassionate, and sometimes analytic and withholding. It is not only the gender but also the circumstances that dictate the behaviour of a person. To be female (or to be male) involves an entire way of viewing the world, of relating to people, of thinking, of communicating. That difference must be taken seriously.

Although Hinduism has become an integral part of our culture, in a way the Hindu religion seems to contradict the culture or vice versa much of the time. However, within the Hindu religion there is no defined authority with recognized jurisdiction. A man, therefore, could neglect any one of the prescribed duties of his group and still be regarded as a good Hindu. In Hinduism, none is therefore regarded as having forsaken his or her religion, even if he or she deviates from the usually accepted doctrines or practices.

Notes

1 Oriya and Hindi belong to the Indo-Aryan language family, while Kannada and Malayalam belong to the Dravidian family.

2 Bride burning is often related to dowry, when the bride's family cannot pay the amount demanded by the in-laws. When the deadline specified for paying runs out, the bride is burned and the incident is often passed off as accident.

3 Beames (1966) claims that Bengali and Oriya have no linguistic gender at all, except in the pure Sanskrit *Tatsamas*, which retain the form of the Sanskrit genders.

4 Priestly (1983) maintains that many Indic languages (e.g., Assamese, Bengali, Nepali, Oriya) have lost gender.

5 Of course, there is a word *patita* 'fallen man', but the meaning associated with it does not denote the sin of extramarital affair committed by man.

6 *Mahaapru* is the contracted form of *mahaaprabhu* 'the great master' or 'god'.

References

Agyekum, K. (1996). Akan verbal taboos in the context of the ethnography of communication. *Working Papers in Linguistics*, No. 30. Trondheim, Norway: Norwegian University of Science and Technology.

Beames, J. (1872–1878, 1966). *A comparative grammar of the modern Aryan languages of India: To wit, Hindi, Punjabi, Sindhi, Gujarati, Marathi, Oriya and Bengali*. Reprinted 1966. Vol. 2, Delhi: Munshiram Manoharlal.

Butler, J. (1990). *Gender trouble: Feminism and the subversion of identity*. London: Routledge.

Clark, H., and Clark, E. (1977). *Psychology and language: An introduction to psycholinguistics*. New York: Harcourt Brace Jovanovich.

Hasan, K. (1997). *Aadimaru aadhunika [From ancient to modern]*. Cuttack: Popular Publications.

Irigaray, L. (1993). *Je, tu, nous: Toward a culture of difference*. (trans. A. Martin). London: Routledge.

Khokle, V. S. (1995). 'Gender and Marathi,' paper presented at the National Seminar/workshop on Language and Gender in Indian Languages, Baroda.

Priestly, T. M. S. (1983). 'On 'drift' in Indo-European gender systems.' *Journal of Indo-European Studies* 11: 339–63.

Sahoo, K. (2003). 'Linguistic and socio-cultural implications of gendered structures in Oriya.' In: M. Hellinger and H. Bushman (Eds), *An international handbook of gender across languages*. Vol. III, Amsterdam: Benjamin Publications. (pp. 239–57).

Index

Note: The notes at the end of chapters have not been indexed.

abbreviations 209
Abdel-Jawad, H.R. 42, 47, 48–9, 50, 54
aboriginal languages 88, 90–1, 92, 95
Abu Hanifah, Imam 120
Abu-Haider, F. 42
academic efficacy 244, 249
address, terms of 50–2
addressee 110, 111–13
addressor 110, 111–13
Adhan 119
Africa 158
African traditional religion 225
age 93–5, 96–8
agency 203, 223–4
Agyekum, K. 265
Alexander, N. 225
aloud reading 155
ambivalence 235
American women: cursing habits and religiosity 63–82
  1986–1996 64–5
  cross-cultural comparisons 65–9
  past and present 63–5
  power and taboo language 76–81
  religion, learning and language restrictions 69–72
  religious personality and cursing 72–4
  sexual anxiety, guilt and repression 74–6
Ames, C. 241
Amsterdam, P. 180, 182
ancestry 78
Anglican Church 134, 135, 137
animal terms 66, 68
anti-sexual group 73
anxiety, sexual 74–6
Aphrodite 19
Aquinas, T. 31
Arabic 88, 99
  see also asymmetries of male/female representation within Arabic
Arishvarar, A. 16
Arluke, A. 77
Asia 158
assertions, strong 215
assertive language 207, 209–13
asymmetries of male/female representation within Arabic 41–61
  avoidance 43–4
  incongruence 44
  morphological gender 44–6
  personal names 46–50
  referential terms 52–9
  shift 44
  titles and terms of address 50–2
asymmetry in lexical meaning 41, 265–7
Ataque de nervios 69
Athena 19, 21
attractiveness 78
audience 110, 111–13
Augustine, Saint 31
Australia 158
Austria 33
avoidance 43–4, 47, 48, 49, 52
  name 43, 46
  negative 53–6
  positive 52, 53–6
awrah 119, 125

Bacon, S.M. 241
Badawi, E.S. 54, 58
Bailey, K. 153
Bainbridge, W.S. 175, 181
Baker, S.C. 241
Bakir, M. 42
Bamba, Sheikh A. 222
'baraka' 225
Barthes, R. 155
Baxter, J. 154
Beames, J. 263
Berg, D. 169–70, 171–80, 182, 183
Berkowitz, N. 5, 240–54

Bernard, E. 32
Bernini 13
Bhimji, F. 5, 203–13
Bible 70
blasphemy 70, 71
body:
  parts 68
  products 66, 68
Boeri Williams, M. 169
Bonvillain, N. 241
Børresen, K. 14
Bourdieu, P. 224
Boyle, J. 241, 247
Bozeman, J.M. 170
Brabant, S. 79
Brahma 17
Brahmanism 15, 34
Brass, M. 21
Braun, K. 81
Brazil 67, 69
Brecht, R.D. 241
Britain 65–6, 158
British Islamic women asserting positions on-line 203–19
  assertive language 209–13
  computer mediated conversations 206–7
  conveying Islamic identities 208–9
  data and methodology 207
  mixed versus single-sex exchange 213–16
  on-line Islamic groups 205–6
  women as agents of knowledge 216–19
Britto, F. 4, 25–36
broadcast language 73–4
Bruner, E. 77
Bucholtz, M. 203
Buckley, G.A. 28
Buddha 17
Buddhism 1, 3, 4, 18–19, 106
  God and gender, overview of 9, 10, 11
  Taiwan 87–8, 89, 90, 91, 92–4, 95–6, 97–8, 99
Bullock, K. 204
Burkent, W. 20, 21
Burstall, C. 241, 247
Busse, C. 153, 158, 163
Butler, J. 257
Butler, R. 241
buttocks 78

Cambodia 99
Campbell, J. 18, 20
Campbell, W. 91
Canada 4, 5, 151–2, 158
Canadian English 66
Canon Law (1983) 32
Caravaggio 13
Cardoso, F. 67
Catholicism 33, 34, 35, 70, 221
  Taiwan 89–90, 95, 96, 98
celebrative occasion 155
censorship 70, 71
Chancellor, J.D. 170, 181
Chastain, K. 241
Chavez, M. 241, 247
Chen, M. 87, 88
Chen, S. 68
Children of God 168–84
  all or nothing/us them 178
  format and illustrations 174–7
  gender 180–2
  'the girl who wouldn't' letter 171–2
  history 169–71
  minimization of self and eros/agape 179–80
  minimum and maximum 177–8
  orality and conversation 172–4
  particular to general 179
choice 203
Christ, P.C. 27, 31, 35
Christianity 1–2, 3, 4, 10, 11, 12–14, 169
  God and gender, overview of 10, 14
  language use and silence as morality: Evangelical theology college 151, 152, 163, 164
  Saudi Arabian English newspaper: letters to the editor 104, 105, 106
  sexuality in ex-gay ministry 187–9, 190, 196, 200
  Taiwan 87, 88, 89–90, 91, 92–3, 94, 95, 97, 98, 99
  see also Judeo-Christian feminist debates

Christle, D. 81
Chung, S.Y. 68
Clark, B. 32
Clark, E. 135, 263
Clark, H. 263
Clatterbaugh, K. 163
closures 208
clothing 224
Cohen, D. 5, 240–54
collaborative language 216
Collier, S. 5, 221–37
Collins, M. 27
Collins, S. 26
colonialism 225, 231
Comiskey, A. 191–3
communities of practice 187–9, 223, 227
compassion 126–7
Concerned Women for America 153
conflict 136–7, 141–7
Confucianism 87, 91
Congregation for Divine Worship and the Discipline of the Sacraments in Rome 34–5
congregational prayer *see* women leading congregational prayer
Connell, R. 157
Conservative community 242–3, 248
Content Analysis and Ethnography of Speaking 227
conversation 172–4
Coontz, S. 153
Cope, G. 13
Coprolalia: (compulsive cursing) 65–6, 67, 68, 69
Corsaro, W.A. 136
Corson, D. 153
Coser, R. 81
Cox, H. 81
Critical Discourse Analysis 227
cross-cultural comparisons 65–9
cult *see* Children of God
culture-specific facts 114–16
cursing *see* American women: cursing habits and religiosity
Cyber-Parish: on-line Episcopal community 133–48
  conflict and power 136–7
  data and methodology 137–9
  gender and religion 134–6
  male/female overall participation 139–40
  male/female participation in conflict as marker of empowerment 141–7
  male/female participation within core group 140–1

Daly, C.B. 27, 31
Daly, N. 81
Daraqutni, Imam 118–19
Daugherty, S. 81
Davies, B. 236
de Oliveria, J. 67
defensiveness 75
Denmark 68, 69
derogation 44, 48, 56, 57–8
Derrida, J. 12
Devi 15, 17
DiBennedetto, T. 133, 136
disaffiliation 209
disagreement, strong 215
discourse styles 206–7
Dixon, J. 135
doctrine of flirty fishing 170, 171, 175, 178, 181
dominance 81
double entendre words 75, 76
Duncan, T. 69
Dweck, C.S. 241
dyslexia 117, 127–8

east Asia 69
Eckert, P. 188, 241
Eder, D. 136
education (literacy) 89–90, 105, 117, 127–8
Egypt 11, 22, 42, 44
  Ancient 9, 19–23
Eid, M. 43–4, 51
Elaide, M. 71
Elliott, E.S. 241
Ellis, R. 241, 247
Emanuel, G. 75
empowerment 141–7
Emswiler, S.N. 26
Emswiler, T.N. 26

English 43, 44, 45, 50, 51, 58, 67, 68, 69
Taiwan 91, 92, 95, 96
Enroth, R. 170
Episcopal Church 4
see also Cyber-Parish: on-line Episcopal community
equal rights 115, 116–18
equal value 115, 116–18, 121, 128–9
'Ethnography of Communnication' 101–3
euphemisms 41, 71–2, 80
Euripides 19, 20
Europe 26, 106
evangelism *see* language use and silence as morality: Evangelical theology college
exclamation marks 216
Executive Yuan 88
expletives 80–1
extraversion 72

Fairclough, N. 191, 227
Falungong 97
family home, what to avoid in 108
Family of Love *see* Children of God
Farwaneh, S. 4, 41–61
Feinberg, T. 65–6
Feisal, King 105
femininity 187, 201
feminism 153, 156
second wave 203
see also Judeo-Christian feminist debates
feminist poststructuralism 221, 222–4
Ferguson, C.A. 43
Ferreira, C.R. 30
fighting words 77–8
Filteau, J. 34
Finley, M. 32
Fiorenza, E.S. 26, 27, 33
Fishman, S.B. 240, 242, 253
Flexner, S. 70
Flores, A. 19
Focus on the Family 153
Fog, R. 68
forms of address 51
Four Noble Truths 18
Frank, A.W. 154, 155–6
Frank, F. 26
Frazier, N. 152
free-association research 75–6
French 51
fresh talk 155
Fussell, P. 80

Gal, S. 240, 241
Galbraith, G. 75
Gallagher, S.K. 157, 162, 163
Gardner, R.C. 241, 247
generation discontinuities in English acquisition 232–5
generic masculines 263–4
genitalia 66, 68, 78
genre 110–11
genteelism 80
Germany 33, 105–6
Gilchrist, M. 34
Gilligan, C. 157
Giotto 13
Giovanni, V. 241, 242, 247
God and gender, overview of in religion 9–23
Ancient Greece, Rome and Egypt 19–23
Buddhism 18–19
Christianity 12–14
Hinduism 15–18
Islam/Muslims 14–15
Goffman, E. 154, 155
Gold, V.R. 34
Goldenberg, N. 26, 31, 32, 33
Goldstein, T. 223
Gole 204
Gombrich, E.H. 13
Goodstein, L. 151
Goodwin, C. 136
Goodwin, M.H. 136, 205, 209, 212, 213
Gornick, V. 25
Graham, S. 4, 5, 133–48
Greece, Ancient 3, 9, 11, 19–23
Greek 10
Greenberg, J. 54
greetings 208
Grenz, S. 152–3
Grey, A. 74

Grimshaw, A. 136
Grosser, G. 74
Guillemette, N. 32
guilt 74–6
Gumperz, J.J. 102

Haddon, G.P. 26, 31, 33
Hadith 55, 60
Haeri, N. 42
Hahn, K. 75
Hakka 88, 90, 91, 92, 95
Halkes, C. 31
Hall, K. 133
Hamilton, H. 138
Hancock, M. 153
harassment 77–8
  at work 81
Hargrave, A. 73, 74
Harré, R. 137, 236
Hasan, K. 266
Heather 133
Hebrew 10, 42, 54
  see also Hebrew language acquisition in Jewish high schools in North America
Hebrew language acquisition in Jewish high schools in North America 240–54
  classroom goal structure, student's academic efficacy and learning situation evaluation 243–4
  language achievement and language learning, effects of gender on 245–7
  learning outcome 244
  participants 243
  procedure 244
  religious affiliation and coeducation/segregation effects 247–52
  theoretical background 241–3
Helen of Troy 19, 20
Heller, M. 230
Hellinger, M. 46
Helmreich, R.L. 241
Henley, N. 79, 81
Hera 19, 21
heritage 249–50
hermaphrodites 122, 125
hermeneutics of suspicion 30
Herring, S. 133, 134, 136, 138–41, 147, 206–7, 212
Herrmann, D. 72–3
Hersh, K. 81
Hindi 258
Hinds, M. 54, 58
Hinduism 1, 3, 5, 34, 106
  God and gender, overview of 9, 11–12, 15–18
  see also Hinduism and the Indian woman
Hinduism and the Indian woman 257–68
  asymmetry in lexical meaning 265–7
  generic masculines 263–4
  selfhood 259–63
  socio-cultural set-up 258–9
  taboo expressions and gender terms of abuse 264–5
Hitchcock, H.H. 34, 35
Hoffman, E. 230
Hole, J. 26
Holland, D. 78–9
Holmes, J. 81, 154, 189, 205, 223
Holmes, L. 241
Homer 19
homosexuality 78
Hong Kong 65–6, 68–9
Hooft, W.A.V. 12
Horus 11, 22
Howard, J. 196
Hsue, C. 99
Huang, S. 90
Huebner, T. 241, 247
Hughes, A.C. 34
Hughes, G. 63
Hughes, S.E. 80–1
Hurcombe, L. 25
Hymes, D. 102, 111–12, 138, 227

I-kuan Tao 90
icons 208, 216
identity 3, 206, 208–9, 217
  see also Senegalese women and identity
Imamah 118
inclusive language 33–5

incongruence 44, 46, 48, 49–50, 58–9
India 106
 see also Hinduism and the Indian woman
Indonesia 99, 114
Industrial Revolution 221
informal language 209
Ingersoll, J. 157, 162
inheritance usurped 107
insults, gender-related 78–9
intensifier adverbs 214
International Commission on English in the Liturgy 34
interviews 92
Iran 48
Iraq 42
Irigaray, L. 169, 176, 180, 181–2, 260
Isis 11, 21, 22
Islam/Muslims 1–2, 4, 9, 10, 14–15, 26, 71, 88
 in Britain 5
 Muridiyya 222
 Saudi Arabian English newspaper: women's letters to the editor 103–4, 105, 106, 114, 116, 122, 124
 Taiwan 91, 99
 see also British Islamic women asserting positions on-line; Qu'ran; Senegalese women and identity through Islamic practice
Italian 51

Jackson, M. 225
Jainism 106
Japan 28, 65–7, 69
Jaworski, A. 157
Jay, T.B. 4, 63–82
Jensen, J. 35
Jerome 31
Jews 5
 see also Hebrew; Judaism
Johnson, D. 133, 136
Johnson, E. 10, 11, 12, 28, 29, 31, 33, 35
joke telling 77, 81
Jordan 42
Jordanian Arabic 49
Judaism 1, 3, 10, 14, 106
 see also Judeo-Christian feminist debates
Judeo-Christian feminist debates 4, 25–36
 feminist alternatives 32–3
 feminist involvement in Church affairs 25–7
 inclusive language 33–5
 male metaphors, feminist objections to 30–2
 male metaphors, traditionalist defence of 29–30
 maleness of the Judeo-Christian God 27–8
Jule, A. 1–6, 151–65
Julian of Norwich 28, 30
Jung, C. 20

Kaczor, C. 27, 29–30
Kalki 17
Kamaljkhani, Z. 204
Kanfer, S. 34
Kannada 258
karma 259
Kaufman, I. 30
Kelly, C. 30
Kelson, J. 77
Kent, S.A. 182
Kerswill, P. 236
Khokle, V.S. 262
King, K. 69
King, U. 17, 182
kinship terms 45
Kjesbo, D.M. 152–3
Klee, C. 241
Kniffka, H. 4, 5, 101–31
knowledge, women as agents of 216–19
Kobliner, H. 244
Kouritzin, S. 235, 236
Koven 236
Kramarae, C. 26, 154
Kutakoff, L. 77

laboratory studies 74
Labov, W. 91

Lacan, J. 156
LaCugna, C.M. 27, 30, 31
Laczek, W. 74
Lakoff, R. 26, 41, 56
Lambert, W.E. 241, 247
Langton, R. 183
language, strong 213, 216
language use and silence as morality: Evangelical theology college 151–65
  classes 159–61
  lecturing as power 156–7
  lecturing as teaching method 153–6
  morality as gendered 157–8
  study 152–3
  theology college 158–9
Lansborough, M. 91
latah 69
Latin 10
Latin America 69, 158
Lave, J. 189
Lawless, E. 134, 135
laxity 73
learning restrictions 69–72
Lectionary of the Mass 34
lecturing 153–7
Lee, P. 68, 133
Lees, A. 65–6, 67, 68
Leggett, E.L. 241
Legman, G. 70
Leiberman, H. 75
Leonard, W. 77
lesbian relationships *see* Children of God
Levi-Strauss, C. 19
Levin, J. 77
Levine, E. 26
lexemes 44
lexical asymmetries 41, 265–7
Liang, A.C. 203
Liao, C. 4, 5, 87–99
liberality 73
Lieh-Mak, F.L. 66, 68
linguistic:
  resistance 235–6
  space 161–2, 163
Lissner, A. 28, 33
Long, R. 72–3
Loving Jesus Revolution 180
Lynch, A. 241

Maalej, Z. 57
McConnell-Ginet, S. 188, 241
McCormick, R.A. 32
McFague, S. 13
McGroaty, M. 230
MacIntyre, P.D. 241
McKay, S. 230, 236
McLaughlin, E. 30
McWilliams, E.M. 138
Maehr, M.L. 241
Mahabharata 18
Mahmood, S. 204
Mahony, P. 153
Malayalam 258
male metaphors:
  feminist objections 30–2
  traditionalist defence 29–30
male/female representation *see* asymmetries of male/female representation
Malik, Imam 119, 125
Mandalam, S.S. 12
Mandarin 88, 90, 92, 95, 96, 99
Manisha 17
Mankowski, P. 27, 34
marginalization 56
Maria 169
Marlowe, M.D. 34
Martell, K. 81
Martin, F. 26
Martin, J.H. 27
Martin-Jones, M. 230
Martines, L. 27
Martyna, W. 31
masculine:
  form before the feminine 41
  generic 263–4
  pronoun 41
masculinity 187, 198–9, 200, 201
mastery 249
media stereotypes 79
medicines while fasting 109
Mendoza-Denton, N. 205, 209, 213
Menelaus 19
Mercy 170

message:
  content 110, 121, 124
  form 110–11, 121, 124
  long 207
metacommunicative comment 123
metaphors 13, 122
  see also male metaphors
Metz, J.B. 31
Meyerhoff, M. 189, 205, 223, 241
Middle East 66
Middleton, M.J. 242, 247
Midgley, C. 242, 244, 247
Millet, K. 25
Milner, J. 75–6
Missal 34
Mitchell, C. 77
Moloney, G. 35
Mooney, A. 4, 5, 79, 168–84
morality as gendered 157–8
Moran, B.K. 25
Mormons 71
morphological:
  features 59–60
  gender 44–6
Morrow, N. 235
Morton, N. 26
Moses, T. 75–6
Mosher Forced-Choice Guilt Scale 75
motivations 96–8
Mourides 225
Muhammad, Prophet 55
Murad, A.H. 10, 11, 14
Mutch, B.H. 152, 153, 157–8

name avoidance 43, 46
neuroticism 72
Newman, B.M. 34
Newton, J. 81
Nicholls, J.G. 241
Nirvana 19
Nolen, S.B. 242, 247
Nomura, Y. 66–7
non-anatomical group 73
Nora, L.M. 81
North America *see* Hebrew language acquisition in Jewish high schools in North America
Norton Pierce, B. 223
Nyikos, M. 241, 247

obituaries 43–4
objectification 57
O'Brien, E. 28
obscenities 64–5, 66, 67, 71, 73, 81
occupation 82
Oddie, W. 29
O'Faolain, J. 27
offended group 73
offendedness 72–3
offensiveness 66, 72–3
on-line religion *see* British Islamic women asserting positions on-line; Cyber-Parish: on-line Episcopal community
oppositional orientations 207, 215
orality 172–4
Orellana, M.F. 205
Orientalism 204
Oriya 258, 263, 264–5, 266–7
Orthodox Christians 25, 242–3, 248, 252, 253–4
Ostling, R.N. 30
others doing wrong 108
overt disagreement 209
Oxford, R. 241, 247

Paglia, C. 151, 164
Pajares, F. 241, 242, 247
Pakistan 114
Pakkenberg, B. 68
Pakkenberg, H. 68
Palestine 50
Panyametheekul, S. 206–7, 212
paradigmatic asymmetries 41
paralellisms: superficial 52–3
parallelisms 123
  superficial 52–3
Parashurama 17
Paris (Prince) 19
parochial school 74
partial repeats 212
participant observation 92
participation:
  patterns 206
  rates 139–41
Parvati 11
Pauwels, A. 41, 44
Pavlenko, A. 236, 240, 241
Payne, L. 193, 195

Pederson, V. 75
Peebles, A. 5, 187–201
pejoration 57
Pennycook, A. 230–1
performance approach 250–2
permissive respondents 73
Persian 43
personal names 46–50
personality, religious 72–4
Philippines 106
Phillips, S. 136
Phoenician 54
Piller, I. 240, 241
Pius XII, Pope 32
Plaskow, J. 27, 31, 35
Politzer, R. 241, 247
Ponticelli, C. 195–6
pornographic illustrations 175–6
Portugese 44
poststructuralist approach 240–1
  see also feminist postructuralism
power 80, 81, 82, 136–7, 223–4, 230, 235
  Children of God 171–2, 174–5, 182
  and lecturing 156–7
  see also power and taboo language
power and taboo language 76–81
  harassment and fighting words 77–8
  insults, gender-related 78–9
  joke telling 77
  jokes and harassment at work 81
  media stereotypes 79
  sexual terminology 77
  working-class women 80–1
Presbyterianism 91
Preston, K. 78
Priestly, T.M.S. 263
profanities 66, 68, 69, 70, 71, 72, 80, 81, 82
Protestantism 91, 151, 152, 164, 221
psychology 189–93
public sphere 204
publicity 73

Qadiriyya 225
questionnaire analysis 92–5
questions 212, 215
Qu'ran (Koran) 2, 70, 99, 122
  asymmetries of male/female representation in Arabic 55, 59, 60
  God and gender, overview of 10, 14, 15
  Saudi Arabian English newspaper: women's letters to the editor 105, 119, 122, 125, 128

racial/ethnic prejudice 106
Ramayana 18
Raming, I. 13
Rampton, B. 231
Ranke-Heinemann, U. 27, 31
Raschke, C. 12
Raudvere, C. 204
receiver 110, 111–13
referential terms 52–9, 60
  avoidance, positive and negative 53–6
  incongruence 58–9
  semantic shifts 56–8
  superficial parallelism 52–3
Reform community 242–3, 248
Regeur, L. 68
religion 66, 73
religious:
  freedom 223–4
  names 47
  police 104
  protectors 73
  studies 236
repression 74–6
research, free-association 75–6
resistance 225
responsivity 75–6
Revised Standard Version of the Bible 34
'Revival, The' website 207
Revolutionary Women 181
rhetorical:
  comment 123
  language 207
  questions 212, 215
Richard, D. 77
Richardson, H. 135
rights and privileges 128–9
  see also equal rights

Risch, B. 78
Rizzo, T.A. 136
Robertson, M. 66
Robinson, J.L. 241
Robinson, W. 77
roles 206
Romaine, S. 43, 44, 56, 58
Rome, Ancient 3, 9, 11, 19–23
Ruether, R.R. 27, 31
Ruth, S. 26, 27, 31

saalah 118–19, 125
Sabalaskey, B. 34
Sadker, D. 153, 154, 242, 253
Sadker, M. 152, 153, 154, 242, 253
Sahih:
  Bukhari 55
  Muslim 55
Sahoo, K. 263, 267
Said, E. 56–7
samsara 19
Sanders, J. 77
Sanskrit 88
Santini, S. 28
Sarah, E. 154
Saudi Arabian English newspaper: letters to the editor 4, 5, 101–31
  corpus 107–10
  'Ethnography of Communnication' 101–3
  socio-cultural setting in 1980s and 1990s 103–6
  see also socio-linguistic analysis
Schill, T. 75
Schillebeeckx, E. 31
Schneider, L. 75
Schneiders, S. 34
Schroeder, M. 133
seaminess 73
Second Language Acquisition 236
second marriages 108
Segawa, M. 66–7
self-empowering identity 217
self-identity 208–9
selfhood 259–63
semantic/semantics 45
  derogation 41
  features 59–60
  shift 44, 46, 48, 56–8
sender 110, 111–13
Senegalese women and identity through Islamic practice 5, 221–37
  community of practice 227
  English, minimal use of 228–30
  feminist poststructuralism: power, agency and religious freedom 223–4
  generation discontinuities in English acquisition 232–5
  linguistic resistance 235–6
  methodology 225–7
  religion and tradition, intersection of 230–2
sensitivity 78
sensitization 75
Serapis (Osiris) 11, 22
sex, right to 172
sexual:
  anxiety 63, 74–6, 82
  graffiti 77
  looseness 78
  terminology 77
sexuality 73, 78
  see also sexuality in ex-gay ministry
sexuality in ex-gay ministry 187–201
  community of practice 187–9
  discourse 193–6
  theology and psychology 189–93
  transforming linguistic practice of gender 196–200
Shabadi, K. 5, 257–68
Shakti 16–17
Shapiro, A. 65–6
Shapiro, E. 65–6
Shariah code of justice 104
Shariah court judge *see* women leading congregational prayer or being Shariah court judge
Sheshar-Aton 242
shifts, semantic 44, 46, 48, 56–8
Shiites 48
Shiue, C. 241
Shiva, Lord 11, 16–17
Siegel, M. 225
Skelton, C. 164
Skinner, D. 78–9
Smith, C.R. 34

Smith, P.M. 26
social appropriateness 206
social class 63, 80–1, 82
socially offensive words 66
socio-cultural set-up 103–6, 258–9
sociolinguistic analysis 110–30
  culture-specific facts 114–16
  equal rights or equal value for women 116–18
  'message form' and 'genre' 110–11
  'sender', 'addressor', 'receiver', 'addressee' and 'audience' 111–13
  see also women leading congregational prayer or being Shariah court judge
Sole, Y. 241
Spain 67–8, 69
Spanish 51
Spence, J.T. 241
Spender, D. 26, 31, 154
Spolsky, B. 236, 241, 242, 247
Stackhouse, J.G. 152
Stanley, K. 78
Stanton, E.C. 26
Starhawk 26, 33
status 82
Steichen, D. 30
stereotypes 57, 199–200
  media 79
Stone, M. 26
Stubbe, M. 81
Sufi 14–15
  Islam 224, 225
Sulla, L.C. 11
Sullivan, G. 81
Sunderland, J. 153
Sunnah 208
Sunnis 105
Sura 14
Sutton, L. 203
Swigart, L. 235
Swindler, A. 26
Switzerland 33
symmetry 53

taboo language 47, 63, 74, 264–5
  see also power and taboo language
Tabouret-Keller, A. 236
Taiwan 5, 87–99
  education and religion 89–90
  interview and participant observation 92
  languages and religions 90–1
  languages used in places of worship 95–6
  motivations 96–8
  questionnaire analysis 92–5
Talbot, M. 57
Tannen, D. 98, 136, 141, 223
Tantric tradition 12, 15
Taoism 87–8, 89, 91, 92–5, 96–8
  I-Kuan 88
Tekcan, M. 3–4, 9–23
terms:
  of abuse, gender 264–5
  of address 50–2, 215
  of reference 53, 54
Terry, R. 77
Tertullian 31
textpragmatic argument behaviour 121, 124, 125, 127
Thailand 99
theology 189–93
Thornborrow, J. 153
Tijaniyya 224
titles 50–2
Tiwanak, G. 81
touchiness 112–13
Tourette Syndrome 65–6, 67, 68, 69
tradition 230–2
Trautman, D.W. 32, 34
Treichler, P. 154
Trimble, M. 66
Tripitaka 19
Turkey 204
Tzu Chi 89, 90

ulema 105
United States 4, 5, 26, 106, 207
  language use and silence as morality: Evangelical theology college 151, 152, 158
  Senegalese women working in 5
Upanishads 18
Urda, J.A. 34
Ushijima, K.C. 71

Vamana 17
van Langenhove, L. 137
Vasquez, A. 27
Vatican 34
Veado, C. 67
Vedas 17–18
Vedic 15
veiling 208
vertical inclusive language 34
Vietnam 99
Vishnu 17
visual cues 172
voice 203

Wachowiak, D. 75
Wahhabites 103
Walker, B.G. 26, 33
Walker, D. 81
Walkerdine, V. 153
Wallis, R. 181
Walsh, A. 74
Walsh, R. 77
Wang, C. 88
Wangerin, R. 169, 174
Weedon, C. 223, 224
Wells, J. 77
Wenger, E. 189, 228–9
Werbner, P. 204
Williams, M. 170, 181
Wilson, L. 19
Wisnesky, R. 34
Wober, J. 73
Wolof nationalism 225
woman's appearance and dress 109
women centre 258
women leading congregational prayer or being Shariah court judge 118–30
  compassion 126–7
  education 127–8
  equal in value 121, 128–9
  message content 121, 124
  message form 121, 124
  metacommunicative comment 123
  metaphors 122
  parallelisms 123
  Qu'ran (Koran) 122
  rhetorical comment 123
  rights and privileges, differences in 128–9
  textpragmatic argument behaviour 121, 124, 125, 127
women society 258
Wong, F. 230, 235, 236
word-association test 75
working-class women 63, 80–1
workplace harassment 81

Yassin, M.A.F. 50
Yisikakafute, Y. 91
Young 65–6

Zentella, A.C. 241
Zeus 21
Zheng-yen, S. 89